Chemistry in Industry & Academia

Volume 1

By

DR. HEBAH ABDEL-WAHAB

Table of Contents

Acknowledgments

The author wishes to acknowledge NJIT Academic and Research Computing Systems resources and staff for computational Facilities and acknowledge the efforts of the Association of Consulting Chemists and Chemical Engineers for making profiles available for our clients in the USA and worldwide.

About The Author

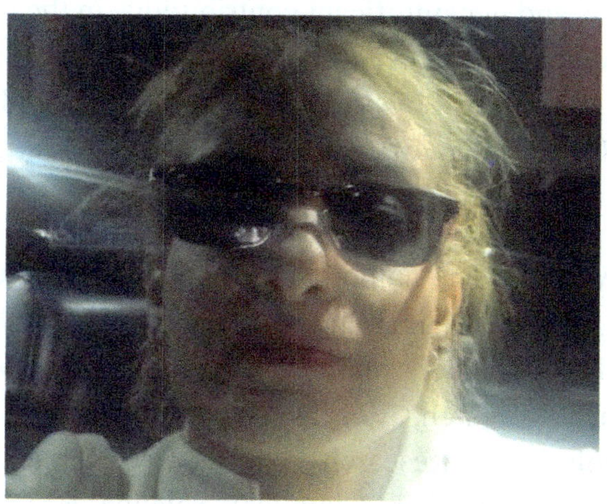

Dr. Hebah Abdel-Wahab

Dr. Abdel-Wahab is a Chemical Consultant at the Association of Consulting Chemists and Chemical Engineers NC, USA, where she is consulted for independent commercial, industrial, and pharmaceutical chemical projects, and an Adjunct Professor of Chemistry at Essex County College & Middlesex County College, NJ, USA, where she has taught chemistry for allied health, preparatory chemistry, general chemistry, organic chemistry, physical chemistry, and inorganic chemistry for 17 years. Professor Abdel-Wahab earned a Ph.D. degree in Chemistry from Rutgers University in 2018. She was scheduled to teach thirty-eight different Chemistry courses during her career: CHM-222, CHM-228, CHM-141, CHM-100, CHE-108, CH- 112, CH-314, CHM-116, CHM-115, CHP-111, CHM-103, Chemistry-307, Chemistry-308, CHEM-4610, CE-241, CHEM-180, CHM-126, CHM-142, CHEM-1105, CHEM- 2208, CHEM- 1033, CHEM-1034, CHEM-2203, CHEM-2204, CHEM-180, CHEM-181, CH-101, CHEM-1117, CHEM-1028, CHEM-2261, CHEM-100, CHEM-111, CHEM-208, CH-132, CHE-101, CHE-102, CHEM-3184, CHEM-1083, and she adapted one-hundred and fifteen different textbook and laboratory manuals.

Dr. Abdel-Wahab was conferred an Honorary Doctor of Science (D.Sc.) degree in Physical Chemistry in December 2022 by the International

Agency for Standards and Ratings for her significant contribution in the field of chemistry, and her significant contributions in the transformation of society. Prof. Abdel-Wahab was also conferred an Honorary Doctor of Science (D.Sc.) degree in Physical Chemistry in December 2022 by the British National University of Queen Mary, WP Road, London, United Kingdom.

Dr. Abdel-Wahab is the author of several book chapters in academic, commercial, industrial, and pharmaceutical chemistry. Prof. Abdel-Wahab is the author of five book chapters published by BP International; Progress in Chemical Science Research, Vol 1, Current Topics on Chemistry and Biochemistry volume 3, Research Aspects in Chemical and Materials Sciences Vol. 2, Progress in Chemical Science Research Vol.4, and Research Aspects in Chemical and Materials Sciences Vol.3, and the author of a book chapter published by Akinik publications; Research Trends in Medical Sciences, Volume 19.

Dr. Abdel-Wahab joined the editorial cabinet as an editorial board member in 2022. She is currently an editorial board member for the Journal of Pharmaceutics and Pharmacology Research, JPPR, 16192 Coastal Highway, Lewes, DE 19958, USA, International Journal of Anesthesiology and Practice, 1603 Capital Avenue, Suite 413-A, Cheyenne, WY, 82001, the USA, and Clinical Case Reports and Trials Medires PUBLISHING LLC, 1603 Capital Avenue, Suite 413-A, Cheyenne, WY, 82001, USA. Dr. Abdel-Wahab is currently a reviewer for Advances in Chemical Engineering and Science (ACES), a SCRIP international Journal, and the Royal Society of Chemistry Journal, RSC Burlington House, Piccadilly, London W1J 0BA.

Dr. Abdel-Wahab belongs to numerous scientific and educational organizations, including the International Society of Female Professionals (ISFP) and the Association of Consulting Chemists and Chemical Engineers (ACC & CE). Professor Abdel-Wahab has been published in Strathmore's Who's Who Worldwide, Farmingdale, NY- 2021, as an Individual Who has Exemplified Leadership & Achievement in their Occupation, Industry, and Profession, and she has also been published in Marquis Who's Who- Berkeley Heights NJ – 2021, as an Individual who Demonstrates

Outstanding Achievement in Their Filed & who has made Innumerable Contributions to Society as Whole.

Preface

The book is divided into ten chapters. Chapters 1-5 discuss methods used to improve the quality of some Pharmaceutical/ industrial products: mouthwashes, acrylic paints, car-wash shampoos, matt and gloss paints, and some bubble-forming tablets. The methods used are based on the varying chemical properties, physical properties, and amounts of the chemical compounds used in the formulation.

Chapter 6 discusses the methods used in thermodynamically identifying the energetically favored products of oxidation of a di-alkyl-thioether, and methods used to calculate thermodynamic parameters, including enthalpy of reactions, entropy of reactions, and Gibbs free energy of reactions. It includes the listing of the thermodynamic table for literature values for enthalpy of reactions, entropy of reactions, and Gibbs free energy of reactions for all possible reaction pathways. It also identifies the energetically favored products for the oxidation reaction.

Chapter 7-9 discusses methods used to calculate kinetic and thermodynamic parameters for simple fluorinated organic alcohols, including mono and di-fluorinated ethanol, muti-fluorinated ethanol, and a di-alkyl-thioether (ethyl methyl sulfide). The methods used to calculate the thermodynamic parameters, including enthalpy of formations, entropy and heat capacity at different temperatures, are discussed. The optimized structures for molecules studied and the optimized structures for possible transition states are included. Tabulated literature values for enthalpy of formations and enthalpy of reactions and literature values for entropy and heat capacity at different temperatures are also included.

Chapter 10 discusses the effect of the obstructed ability to breath by a physical block limit or by prolonged hypoventilation, or by a pulmonary condition on hypoxia, hypercapnia, and blood chemical PH imbalance. Mechanism of alternating blood PH values, methods of identifying blood chemical PH imbalance, methods of measuring P_{CO2}, P_{O2}, and blood PH, types of joint diseases caused due to blood chemical PH imbalance, current medicine prescribed in their treatments, and the natural treatments of such joint diseases are discussed.

This book is based on the actual solutions to industrial problem requests submitted to me through the Association of Consulting Chemists and Chemical Engineers, Cary, NC, USA, and the actual results of thermodynamic and kinetic independent research studies at NJIT, Newark, NJ, USA, and published articles. I hope you find the information in this book helpful to your business or project, and I also hope the information in this book increases your wealth of knowledge.

Dr. Hebah Abdel-Wahab

Edison NJ, USA

chemistryinindustryandacademia@gmail.com

Chapter 1
Chemical Formulations for Mouthwashes

1. Overview

Mouthwash can strengthen the teeth, prevent gum disease, kill bacteria in the mouth that causes gingivitis and improve your oral health. This chapter discusses the method used to improve the properties of the current mouthwash formulation, determine the chemical compounds and safe amounts to be used from the current formulation: distilled water, sodium chlorite, castor oil, peppermint extract, lemon juice directly from the fruit, coconut oil, sodium benzoate, citric acid, and tetrasodium EDTA, and recommend a suitable preservative for the current formulation.

2. Introduction

An independent business owner in New Jersey is looking to produce a small amount of homemade mouthwash to be sold and distributed locally. The ingredient included the following chemicals: distilled water, sodium chlorite, castor oil, peppermint extract, and lemon juice directly from the fruit, coconut oil, sodium benzoate, citric acid, and tetrasodium EDTA. The chemical compounds to be used in the formulation and the safe amounts of each chemical in the formulation are identified.

Mouthwash can strengthen the teeth, prevent gum disease, kill bacteria in the mouth that causes gingivitis and improve your oral health. If the mouthwash contains fluoride, it reduces the cavity when used correctly. It is recommended by dentists to use mouthwash daily in addition to daily brushing and flossing. People with medical conditions, tooth sensitivity, dry sockets, and Xerostomia find mouthwash essential in their daily routine.

Mouthwash is used to refresh and clean the oral cavity. To prevent the development of caries, a formulation containing fluorides has been used, as it destroys and inhibits the oral microbial population that generates malodors or dental plaque. Mouthwash formulation mainly consists of ethanol, water, a humectant, a surfactant, flavor, color, and an active ingredient. Ethanol is used to enhance the flavor impact and adds freshness. Humectant such as glycerin, sorbitol, or propylene glycol adds body to the formulation and improves the mouth feel when using the mouthwash. The surfactant is usually a nonionic such as poly oxy-propylene or poly oxyethylene co polymers; it produces foams, dissolves the flavor oil, and removes debris. The flavor is an oil used to make the mouthwash pleasant to use. The mouthwash may include an active ingredient, for example, an anti-cares effect or antibacterial effect.

Dental Caries is a multifactorial disease produced within the dental plaque by microorganisms. It involves a process of enamel demineralization and remineralization due to the act of organic acids produced within the dental plaque by microorganisms. Enamel demineralization is the process of dissolution of hydroxyapatite by the act of extrinsic or intrinsic acids, according to equation 1, leading to erosion or dental cares. Dental Cares are caused by acetic or lactic acid that diffuses into the enamel pores through the plaque, decreasing the PH of the fluid surrounding the enamel crystals. To prevent dental cares, fluorides, calcium, and phosphates, and antimicrobials can be used as additives in the mouthwash formulation. The presence of fluorides in the microenvironment or around the teeth inhibits demineralization and promotes the remineralization of the tooth surface. The incorporation of fluoride as fluorapatite into enamel decreases its solubility. Casein phospho-peptide amorphous calcium phosphate, CPP-ACP, has been added to mouthwash, tropical pastes, and chewing gums to increase remineralization and decrease demineralization. Antibacterial has some role in caries prevention, it helps with plaque and microbial control, Chloro-hexadiene and triclosan-containing gels, toothpastes, and rinses have been used.

$$Ca_{10}(PO_4)_6(OH)_2 + 2H^+ \rightarrow 10Ca^{2+} + 6PO_4^{3-} + 2H_2O \quad \text{Equation 1.}$$

Mouthwash can be classified into two main types, therapeutic and cosmetic. Therapeutic mouthwash reduce plaque, bad breath, tooth decay, and gingivitis. Children under the age of 6 shouldn't be using mouthwash as they may inadvertently swallow large amounts of liquids unless directed by a dentist. Cosmetic mouthwash has no biological or chemical application beyond its temporary relief, they temporarily control bad breath and gives a pleasant taste and smell.

Active ingredients used in a therapeutic mouthwash are peroxides, fluorides, chlorhexidine, essential oils, and cetyl-pyridinium chlorides. Peroxides are used in whitening mouthwash. Fluorides are used to prevent tooth decay. Chlorhexidine and essential oils are used to fight gingivitis and plaque. Cetyl-pyridinium chlorides is used to prevent bad breath.

Antimicrobial, Fluoride, Anesthetic, and Chlorhexidine mouthwashes are examples of therapeutic mouthwashes.

Antimicrobial mouthwashes are used in the treatment of Oral Malodor and bad breath. Volatile sulfur compounds (VSC's) is the major factor causing bad breath; they arise from the breakdown of food, bacteria associated with dental disease, and dental plaque. Cosmetic mouthwash can also temporarily mask bad breath, but it doesn't kill the bacteria that causes bad breath and has no effect on volatile sulfur compounds causing bad breath. Therapeutic mouthwash containing an antimicrobial is more long-term and can control bad breath. Antimicrobials in mouthwash formulations can include chlorhexidine, cetyl-pyridinium chloride, essential oils (e.g., methyl salicylate, eucalyptol, thymol, and menthol), and chlorine dioxide. Antimicrobial mouthwashes are also used in the treatment of gingivitis and plaque.

Fluoride mouthwashes are used to prevent tooth decay in children and can also be used in the treatment of xerostomia, a condition where the amount of saliva bathing the oral mucous membrane is reduced. It is found that rinsing weekly with 0.2% NaF mouthwash or daily with a 0.02% NaF would reduce dental caries by 20-40% in children.

Anesthetic mouthwashes are used to relieve pain, and they contain anesthetics, dyclonine hydrochloride, phenol, lidocaine, benzocaine, tetracaine hydrochloride, or butamin.

Whitening mouthwash mostly contains 1.5-2% hydrogen peroxide or 10% carbamide peroxide.

Chlorhexidine mouthwash, without the use of antibiotics, is found to be effective for AO prevention following extractions. Dry socket, alveolar osteitis, which is a condition that might follow an extraction procedure, and intense pain 2-3 days after the procedure.

Additives that make a therapeutic mouthwash can be classified into either antibacterial agents or compounds that effect VSC formation. Antibacterial agents help reduce the number of anaerobic bacteria in the mouth. Examples of antibacterial agents that are used in mouthwash formulation are cetyl pyridinium chloride, essential oils/antiseptic formulations, triclosan, and Chlorhexidine gluconate. It is found that the use of 0.75% Cetyl pyridinium chloride (CPC) mouth rinse reduces plaque by 35% over a 6-month period, the use of a 0.2% chlorhexidine gluconate oral rinse solution cause reduction in VSC level (It shows some side effects, it alters the sense of taste and causes staining of mouth and tongue).

Compounds that effect VSC formation neutralize VSC's to improve quality of breath. Examples of ingredients that effect VSC's are zinc compounds and chlorine dioxide (sodium chlorite is an oxidizing agent and is able to degrade sulfur-containing amino acids, which is the building block of VSC's, making fewer of them available to bacteria, leading to fewer VSC's formed). It's found that using mouthwash containing 1% sodium chlorite, reduces VSC for a period of 8h. Some metal ions have the ability to oxidize the thiol, sulfur-based molecules found on VSC's. Examples of metal ions are zinc, copper, magnesium, tin, and sodium; zinc also has an antibacterial property, doesn't cause tooth staining like other metal ions, and has low toxicity. A 1% zinc acetate mouthwash has a significant effect on VSC levels, even 3 hours after use. A typical formulation for essential oils/antiseptic mouthwash is: 0.092% eucalyptol, 0.042% menthol, 0.06% methyl salicylate, and 0.064% thymol in a solution containing 21.6% to 26.9% alcohol, high concentration of alcohol in the formulation (25% ethanol) may cause drying effect on oral tissue, increasing breath malodor. The most effective mouthwash for bad breath is found to use a multifaceted approach, using a combination of formulations. Other products used for breath control may include gum, toothpaste, mints, drops, lozenges, and sprays. It is found that six compounds are mainly used in the formulation of mouthwash designed to fight bad breath (halitosis); zinc chloride, triclosan, essential oils/antiseptic formulations, chlorhexidine gluconate, cetyl-pyridinium, and chlorine dioxide. Studies show that using a 0.09% zinc chloride mouth rinse is effective in reducing the formation of calculus in people.

Sodium chlorite is the main ingredient used in the newer classes of mouthwashes such as ProFresh, CloSYS, TheraBreath, and Oxyfresh. It's sometimes used as a water purifier. These mouthwashes are said to freshen breath up to six hours. SmartMouth uses sodium chlorite as a main ingredient, but it must be mixed with zinc chloride right before use. The zinc ions block receptor sites of amino acids after eating food containing amino acids, so the bacteria can't produce rancid gases, and it lasts for up to twelve hours before another rinse is needed.

Flavors are one of the main ingredients that make a mouthwash, as it gives flavor to the formulation and makes the mouthwash pleasant to use. The different known flavors that can be used in a mouthwash ingredient are: Clove Leaf Oil, Cubeb Oil, Cedarwood Oil, Eucalyptus Oil, Lemon Oil, Italian Type, Coldpressed, Myrrh Oil, Mentha Arvensis Oil, sweet Orange Oil, Peppermint Oil, Sucralose, Saccharin Sodium, Dihydrate, Granular, Spearmint Oil, Camphor, Methyl Salicylate, and Cinnamon Extract.

3. Experimental

The chemical composition of the ingredients used in the making of the mouthwash is identified in table 1.

Distilled water (purified water) is used as a solvent in which the ingredients are soluble in. It's a cooled boiling water steam returned to its liquid state after removing 99.9% of the minerals dissolved in water.

Sodium Chlorite is used as an antimicrobial agent in mouthwash. It reduces bacteria living in the mouth and causing bad breath. It also degrades building blocks of the volatile sulfur compounds causing bad breath, and it degrades sulfur-based amino acids, causing fewer of them being available to bacteria, and fewer volatile sulfur compounds are formed. It is found that one-time use of mouthwash containing 1% sodium chlorite is effective in reducing the RVS level for eight hours and beyond.

Castor oil has the ability to break biofilm, a protective coating created by detrimental bacteria in the microbiome. It makes the environment more hospital to healthy bacteria and eradicates the bad bacteria. It also reduces inflammation of the gums.25. Castor oil is a slightly toxic chemical, and a safe amount of Castor oil is found to be below 5g/kg, below mass percent (m/m) of 0.099%.

Peppermint extract ete is used as a flavoring agent, antiseptic and anti-viral agent. There is no evidence that it has the ability to treat any medical condition. However, the high concentration of menthol in the extract causes side effects in children and infants when nihilated.

Sodium Benzoate is used as a preservative in food, medicine, and cosmetics. It's converted to benzoic acid under these conditions, which are fungistatic and bacteriostatic. Due to the insolubility of benzoic acid in the water, it isn't used directly as a preservative. It's concentration as a preservative is determined by the FDA to be 0.1%.

Lemon juice is a good source of citric acid, containing 1.44g/oz citric acid, which is 4.83 percent by mass citric acid in lemon juice.29. Citric acid is known to cause tooth enamel to dissolve quickly. More dentin is exposed as enamel dissolves, causing teeth cracks and chips, and the edges of the teeth to become more irregular.

Tetrasodium EDTA is a water-soluble acid with chelating properties and is an emulsion stabilizer. It bonds with metal ions in solution, causing them to be inactive, and is used as a preservative for cosmetic formulation and skin care products, creams, lotions …etc. It prevents the change in PH, texture, and color of skincare products. And is used as a co-preservative in skincare formulations. Also, it can enhance foaming and cleaning ability when binding to iron, calcium, or magnesium.

Coconut oil contains 80-90% unsaturated fat, is 100%fat, and contains trace amounts of minerals and vitamins. It is used for cooking food, as a fuel source in industry, and as a base ingredient for the manufacture of soap.

Lemongrass oil is known to be used as an herbal alternative (0.25%) to chlorhexidine (0.2%) in mouthwash. It prevents and treats periodontal diseases.

Table1. Structures, Molecular Formulas, and Molecular weights of all Household Chemical Compounds used in The Current Formulation.

Household Chemical and Chemical Name	Chemical structure	Molecular Formula	Molecular Weight (g/mol)
Distilled water		H_2O	18.015
Sodium chlorite.		$NaClO_2$	90.44

Castor oil.	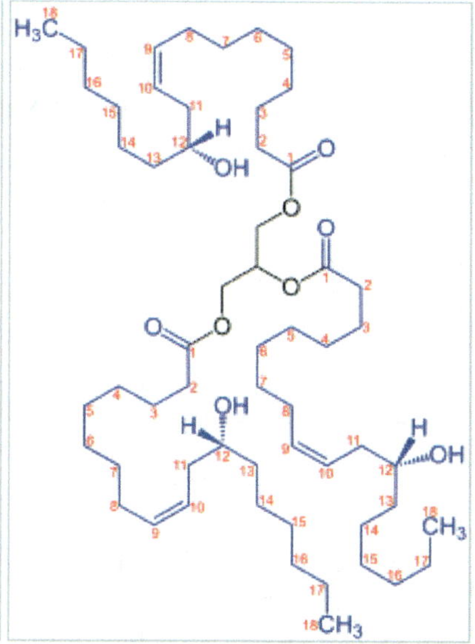 Major component of castor oil, triester of glycerol and ricinoleic acid.	C57H104O9	933.4
Peppermint extract/ Oil.	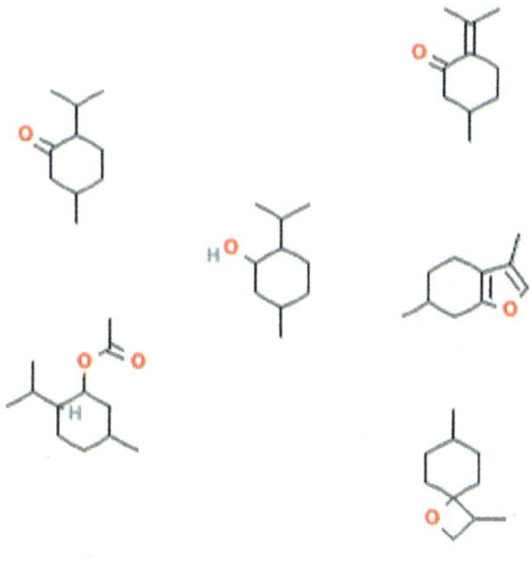 (Menthol), (3,7-Dimethyl-1-oxaspiro[3.5]nonane), (Menthyl acetate), (p-Menth-4(8)-en-3-one), (p-Menthan-3-one) and (Menthofuran).	C62H108O7	965.5

Sodium benzoate.	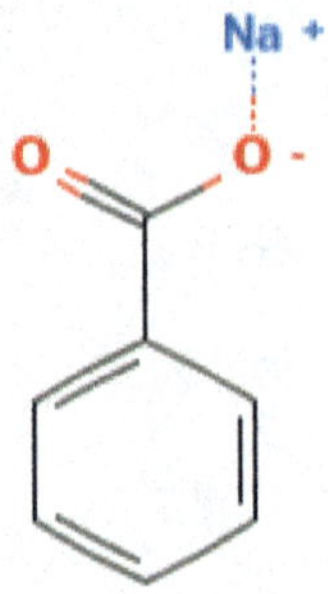	C7H5NaO2	144.11
Lemon Juice (citric acid).	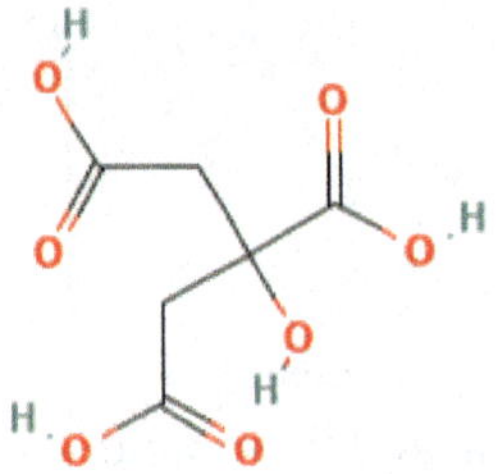	$C_6H_8O_7$	192.124
Tetrasodium EDTA	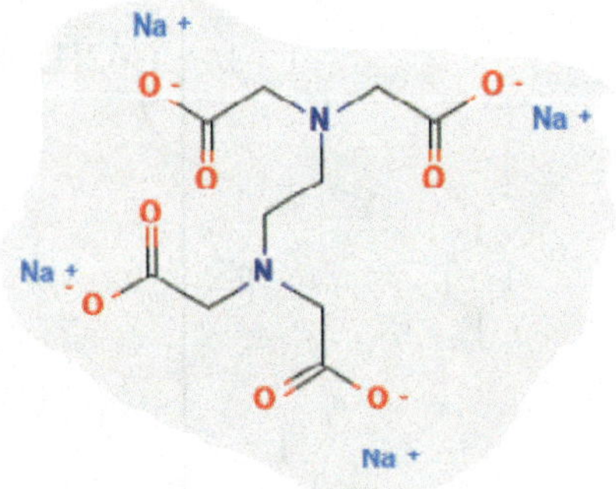	C10H12N2Na4O8	380.171

| Lemongrass oil. | | C51H84O5 | 777.2 |

(Methyleugenol), (L-Borneol), (Nerol), (Dihydromyrcene) and (Neral).

4. Conclusion

The therapeutic mouthwash formulation shouldn't include the following chemical compounds: citric acid, lemon juice, EDTA, and coconut oil. Citric acid/lemon juice is known to weaken and dissolve tooth enamel, and both EDTA and coconut oil are known neither to be used in the pharmaceutical industry nor in mouthwash formulations. EDTA is commonly used in skin products as a chelating agent, and coconut oil is a fat that is known to be used in cooking, fuel, and in the making of soaps.

The therapeutic mouthwash formulation may include 1.0% Sodium chlorite as an active ingredient, 0.1% sodium benzoate as a preservative, water as the solvent, and less than 0.09% percent by mass castor oil. To improve the quality of the mouthwash formulation, other chemical compounds may be used, glycerin as a humectant and poly oxy-propylene co-polymer as a surfactant. Color and flavor may be used to improve the quality of the mouthwash. Some other additives, such as zinc chloride, may also be used; it's known to freshen breath up to twelve times higher when combined with sodium chlorite.

The recommended preservative for mouthwash is sodium benzoate. The safe amount of hydrogenated castor oil is below 5g/ kg.

Chapter 2
Chemical Formulations for Acrylic Matt and Acrylic Gloss Paints

1. Overview

Paints are used to prolong the life of natural and synthetic materials and to protect it; it acts as a barrier against environmental conditions. Mainly paints are made up of solvent, binder, extenders, pigments(s), and some additives. A solvent can be either organic or Inorganic, and it's used as a medium to dissolve paint contents together and make them uniform. It can also be used as a thinner. Pigments are used to give color to the paint. Binders are used to hold pigments in place. Extenders have large pigment particles to improve the adhesion properties of the paint, and they are also used to strengthen the film and save the binder. This chapter discusses the methods used to improve the properties of acrylic matt paint and methods used to increase the scrub resistance test of the current formulation. This chapter also presents the types of additives and chemical compounds that would potentially increase the quality and increase the scrub resistance test for acrylic paints.

2. Introduction

Professional Specialty Chemicals Factory, a polymer production unit in Saudi Arabia, are looking to improve its formulation for acrylic paints to give a better scrub resistance test after decades of using the same formulation. The current formulation consists of the following chemical compounds, 0.4% Natrosol 250- HHBR, 44.39% water, 0.14% Sodium Bicarbonate, 2.0% Polyoxyethylene 25 octyl phenol, 2.0% Octyl phenol polyglycol ether sulfate sodium salt, 0.5%, Provichem, 2.0% Buyl acrylate, 47% Vinyl acetate, 0.1258% Potassium Persulfate,0.0234% Tertiary butyl hydrogen peroxide, 0.06% Hydrogen peroxide, 0.05% Sodium formaldehyde sulfoxylate, 1.0% Di butyl phthalate, 0.1% Formaline, 0.05% *biocide* , 0.1% Silquest A-171 and 0.06% defoamer. The scrub resistance test of the current formulation shows 60% effectiveness compared to the reference; the coat breaks and gloss change were observed after 60 cycles for the current formulation compared to 100 cycles for the reference.

Paints are used to prolong the life of natural and synthetic materials and to protect it; it acts as a barrier against environmental conditions. Paints contain pigment(s), binder, extender, solvent, and some additives. The pigment is used to give color and opacity. Binders are matrices and are used to hold the pigment in place. Extenders have larger pigment particles to improve adhesion and to strengthen the film, and save the binder. A solvent can be either an organic solvent or water and is used as a thinner and to dissolve paint components and make it uniform. Additives are commonly used to improve the properties of the paint. *Contents of an acrylic white matt emulsion paint is known to be as follows:* Pigments 25%, Extender pigments 12%, Additive 5%, Solvent 44%, and Binder 14%.

To control paint gloss, extender pigments are used. Extender pigments give extra weight to the paint and are low-cost pigments. There are two types of pigments, extender pigments, and prime pigments. Prime pigments are those that contribute to both dry and wet hide in paint. There are two main types of prime pigments, inorganic and organic prime pigments. Inorganic prime pigments are earthy colors duller and more durable for exterior paint application. Examples are yellow ochre, umber, and red oxide. TiO2, Titanium dioxide provides the actual color within the container. Organic prime pigments aren't very durable for exterior paint application and provide brighter colors; examples are phthalo blue and hansa yellow. An extender pigment is a naturally occurring chemical substance (usually having a white color) that is added to paint or coating to improve its properties, such as durability, cost, and resistance to corrosion or wear.

Ingredients that provide a binding effect that holds the pigments together as dry film after the liquid solvent evaporates are called binders. Paint binders directly relate to paint performance, including gloss retention or fade resistance, washability, adhesion, and scrub resistance. The binder in many emulsion paints is based on homopolymers or co-polymers of vinyl acetate and a propenoate (acrylic) ester. Other acrylic esters used as co-monomers with ethenyl ethanoate are ethyl propenoate, butyl propenoates, or a co-polymer of butyl propenoate and methyl 2-methylpropenoate. Many commercial extenders are glycerin-based. In fact, glycerin is already an ingredient in most acrylic paints. It gives the paint viscosity in addition to extending the drying time. You can heighten this effect by using glycerol mixed with water. Propylene glycol is also used for thicker applications.

Binders in various emulsion paints are based on co-polymers or homopolymers of a propenoate (acrylic) ester and ethenyl ethanoate (vinyl acetate)

$$CH_3-C\overset{O}{\underset{OH}{\Big\langle}} \;+\; H_2C{=}CH_2 \;+\; \tfrac{1}{2}O_2 \;\xrightarrow{\text{catalyst}}\; CH_3-C\overset{O}{\underset{\underset{H_2C=CH}{O}}{\Big\langle}} \;+\; H_2O$$

ethenyl ethanoate

Solvents are liquids that allow the paint to be applied to the surface directly from the containers. Solvents allow the uniform combination of binders, considered solids, and pigments with the liquid solvent. Types of liquid solvents depend on the type of paint. For Alkyd paints, the liquid solvent is thinner and is typically another type of solvent; for Latex paints, water is the main liquid solvent. More solids in the ingredients allow for higher quality paints, 35-40%. How much paint will remain on the surface after the liquid solvent is evaporated depends on the amount of solid in the paint formulation. The make-up of the solids effects the overall paint performance. High solids by volume don't always give a high-quality product.

To modify the properties of paint and create additional performance properties, additives are used. The percent of additives in the formulation is known to be (01-1%). Additives may include algaecides, bactericides, anti-settling agents, driers, thixotropic agents, dispersants, silicones, and crosslinkers. The molar mass of a polymer is greatly increased by using a crosslinker, silicates, causing the polymer to become a three-dimensional molecule and form a hard film to become resistant to chemical and environmental conditions. Algaecides are used to protect exterior paint films from being disfigured from lichen, algae, and molds, and bactericides are used to preserve water-based paints in their containers. To prevent pigment settling, anti-settling agents (surfactants) may be used. To accelerate drying time, driers are used. Thixotropic agents are known to be used to create a jelly-like texture of the paint that would break down to liquid when the paintbrush is placed in it or when stirred. To improve weather resistance, silicones are used. To stabilize and separate pigment particles and prevent them from aggregating, dispersants are used.

Water-borne emulsion paints are better for the environment than paints made with organic liquid solvents. Emulsion paints are made using emulsion polymerization, and monomers are dispersed in water as an emulsion. The molecular weight of the polymers produced is 500, 000 – 1,000, 000 a.m.u. For a better application of paint, a solvent thinner can be used, and it's either water or organic solvent.

An ideal paint has high scratch and heat resistance, high color stability against ultraviolet and visible light, high water and corrosion resistance, high opacity, a relatively quick drying time and is easily applied to the surface, has high durability and flexibility, has a good flow out of the application and forms a strong protective film.

Surfactants, surface active agents, are compounds that lower the surface tension of liquid or interfacial tension between solid and liquid making them more soluble. Stabilizing effects for latex paints are caused by using a soluble surfactant, as they can form micelles and aggregate structures in solutions. In paints, surfactants are used as an emulsifier. Surfactants have a positive and negative role in being used in a paint formulation. The positive effect would be stabilizing the dispersion of polymer molecules during emulsion polymerization, improving the mechanical stability of paints, and allowing the paint to coat the surface more easily. The negative effect would be decreasing the resistance of the coating as it can be washed out of the coating and enter the environment; some of these surfactants are toxic to the environment, problems with adhesion, and loss of optical clarity. Other chemicals in paint can alter the overall effect of surfactants on paint. The amount of TiO_2 affects the elasticity in latex paint. Chains that consist of alternating silicon and oxygen atoms, and siloxane chains with siloxane tails in polymers have been found to resist hydrolysis and prevent the breakdown of polymer chains; breakdown of polymer chains can cause cracking in the paint film. Siloxane is used in soaps, cosmetics, defoamers, and deodorants.

Figure 2. (a) and (b) are the General formula for a surfactant.

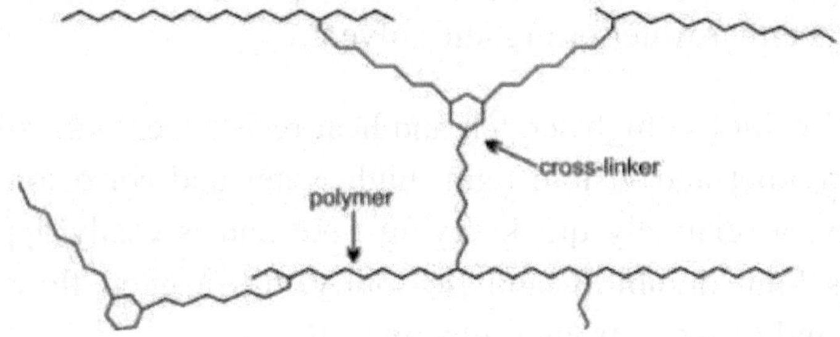

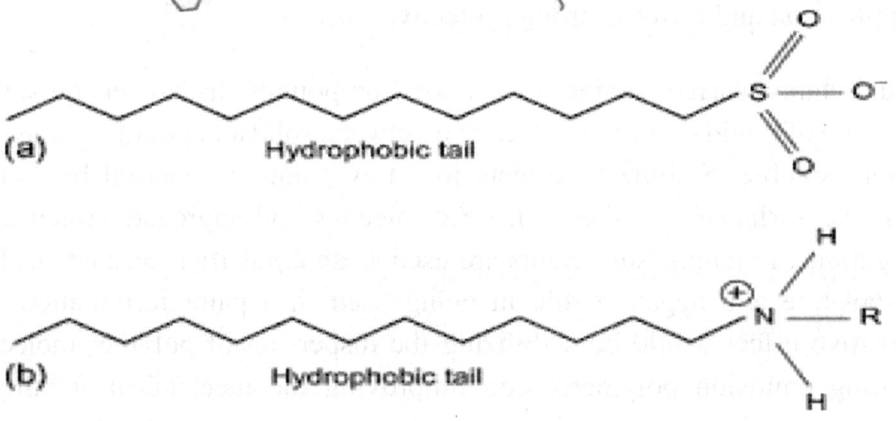

Figure 3. A Crosslinker.

Polymers are high molecular substances that are made up of small low molecular weight chemical species held together by a chemical bond called monomers. Polymers are used in coatings, polyurethane, and plastics, and they are used as a binder in paints. The first polymer studied by chemists is natural or biological polymers such as protein, starch, and cellulose. An example of a polymer would be polyethylene consisting of ethylene monomers that are chemically bonded.

Polyethylene is a polymer consisting of thousands of ethylene units. The name polyethylene is a name arousing from the prefix poly, which means 'many', and the monomer name (ethylene). Synthetic polymers can be either condensation or addition polymers based on the method used in synthesizing them. In addition to polymerization, many molecules are linked together by an addition reaction. The monomer molecules must be unsaturated, containing multiple bonds that can undergo addition reactions. In a free radical addition, an initiator is used to prepare an addition polymer. An initiator is a compound that can produce free radicals, organic compounds with O-O group (peroxides) are commonly used as initiators. When organic peroxides are heated RO-OR, the O-O bond breaks and free radical form (species having unpaired electrons). Alkene monomers would react with the free radical formed and produce another free radical, which in turn reacts with another monomer molecule and produce another free radical molecule in a chain reaction. The reaction terminates when a free radical at the end of the chain reacts with another free radical. In condensation polymerization, many monomer molecules are linked together, and a polymer is formed by a condensation reaction, where a water molecule is removed by joining two molecules. Examples of condensation polymers are polyesters and polyamides. The repeating monomer units are joined together by an ester group in polyesters and by an amide group in polyamides.

Figure 4. Shown is a piece of polyethylene.

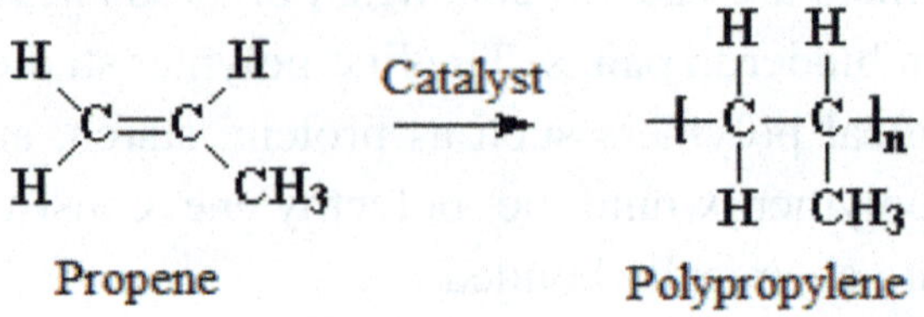

Figure 5. Example for addition polymerization of propene into polypropylene.

$$RO-OR \rightarrow RO\cdot + \cdot OR$$

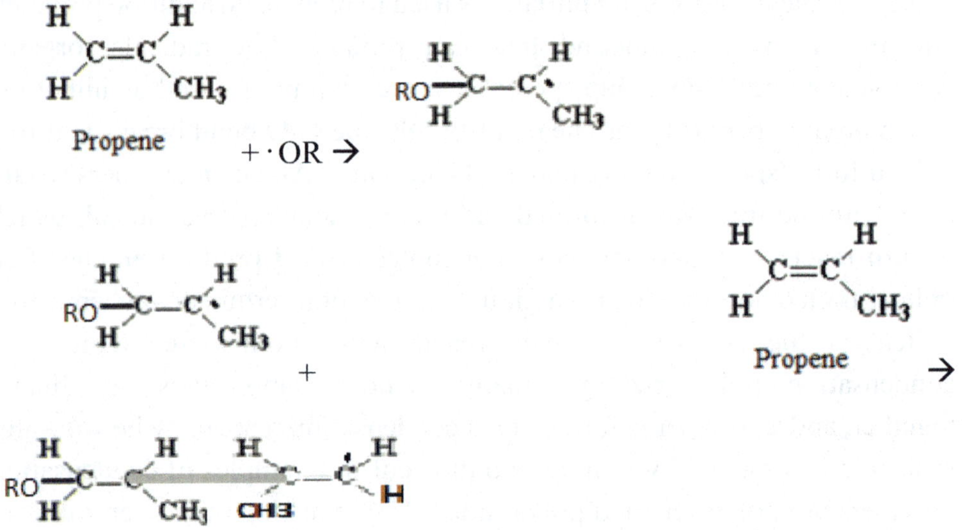

Figure 6. Example for free radical polymerization of propene using organic peroxide.

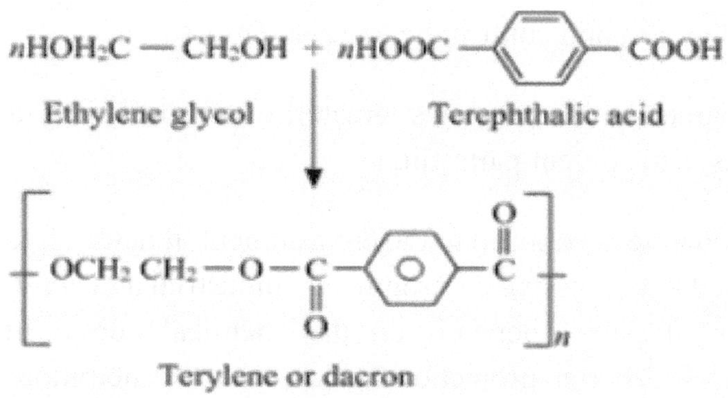

$$nHOH_2C — CH_2OH + nHOOC —⟨○⟩— COOH$$

Ethylene glycol   Terephthalic acid

$$\left[OCH_2\,CH_2—O—\underset{O}{\overset{O}{C}}—⟨○⟩—\overset{O}{C} \right]_n$$

Terylene or dacron

Figure 7. Example of polyester Dacron formed from condensation polymerization of ethylene glycol and terephthalic acid.

The technological and physicomechanical properties of various polymers are dependent on the branching that occurs in the polymer chain compared to a linear polymer. An increased degree of branching in the polymer leads to a decrease in strength properties. For example, branched polybutadiene, divinyl styrene, and other branched elastomers exhibit low elasticity and breaking strength.

Polymers are classified based on their composition into homopolymer and copolymer. A homopolymer is a polymer that consists of only one monomer. A copolymer can be either alternate, random, or block copolymer. An alternate copolymer consists of alternating monomers ab-ab-ab , a block copolymer consists mainly of two or more monomer chains combined, aaaaa-bbbbb, and a random copolymer consists of monomers combined in a random array aa-bb-bb-ab-ba.

The durability of concrete structures can be increased by the application of protective coatings, such as paints.

Protective coatings can be used for the modernization of older concrete works, rehabilitation of deteriorated concrete, and for restoration of its original properties. The chemical composition of the protective coatings determines the film properties. Its properties can be modified by altering the amounts of additives and pigments. Pigments affect the permeability and strength of the film and control the color and gloss.

Latex paints have ten times greater water diffusion, and it's less homogenous than solvent paint films.

When concrete is exposed to the environment, it deteriorates, especially when exposed to aggressive environments: industrial and coastal areas. The performance of paints depends on the chemical composition of their formulation. The better protection against water penetration was greater with increasing the amounts of binders (resin) and decreasing amounts of PVC (poly-vinyl chloride) polymer. In case of concrete with low compressive strength, paints are shown to be efficient for its protection, while high compressive strength concrete paints have smaller importance as it already has a low permeability. Paints are considered for visual aspects with concrete of high compressive strength. Ref actual paper. Decreasing amounts of PVC in the paint formulation from 75-0% showed a decrease in water absorption and an increase in coating effectiveness from 6-91%.

3. Experimental

The paint is prepared by emulsion polymerization, where soap, free radical catalysts (potassium persulfate, peroxides), and monomers are emulsified in water. The water-soluble free radical catalyst is added to induce polymerization.

Natrosol 250- HHBR is used as a nonionic water thickener, a protective colloid, suspending agent, and stabilizer and gives chemical mechanical stability and controls rheology after, during, and before application.

Water is used as a dispensing solvent for emulsion polymerization.

Sodium Bicarbonate is used as a pigment, and it dissolves immediately upon addition to the reaction mixture; a thick textured effect is the result. If more baking soda is used, it causes the paint to fluff up and become more textured.

Polyoxyethylene 25 octyl phenol is used a spermaticide. Octyl phenol polyglycol ether sulfate sodium salt is used as a surfactant and a buffer in emulsion polymerization, copolymerization, and homo-polymerization of monomers, and the sodium salt is used to main PH 6.5-7.5.

Octylphenol Polyglycol Ether Sulphate Sodium Salt is used in the emulsion polymerization of acrylate and methacrylate esters, styrene, and vinyl esters. Sodium Salt can be used in homo-polymerization as well as in copolymerization of these monomers; the sodium salt maintains a pH level ranging from 6.5 – 7.5.

Provichem, Sodium vinyl sulfonate is a surfactant in emulsion polymerization; it produces a copolymer with high water stability.

Butyl acrylate is used as a monomer in copolymerization.

Vinyl acetate is used as the other monomer in homo-polymerization or copolymerization.

Potassium Persulfate is used as an initiator for free radical polymerization and is also known to be used as a bleaching agent.

Tertiary butyl hydrogen peroxide is commonly used as an oxidizing catalyst.

Hydrogen peroxide is known to be used as an initiator in vinyl polymerization in a homogeneous system.

Table 1, lists the chemical name, molecular formula and molecular weight of chemical compounds used in the current acrylic paint formulation.

Chemical Name	Percent (%) used in the current formulation	Molecular Weight (g/mol)	Chemical Structure	Molecular Formula
Water	44.09	18.902	H–O–H	H_2O
Natrosol 250-HHBR	0.4	806.9	N/A	$C_{36}H_{70}O_{19}$
Sodium Bicarbonate	0.14	84.007	HO–C(=O)–ONa	$NaHCO_3$
Polyoxyethylene (2 5) octyl phenyl ether	2.0	250.379	N/A	(C2-H4-O)mult-C14H22-O
Octyl phenol polyglycol ether sulfate sodium salt	2.0	Buffer and Surfactant	N/A	Octyl phenol polyglycol ether sulfate sodium salt

Provichem (*Sodium vinylsulfonate*)	0.5	130.10		$C_2H_3NaO_3S$
Buyl acrylate	2.0	128.17		$C_7H_{12}O_2$
Vinyl acetate	47.0	86.09		$C_4H_6O_2$
Potassium Persulfate	0.1258	270.33		
Tertiary butyl hgrogen peroxide	0.0234	90.12		$C_4H_{10}O_2$

Hydrogen peroxide	0.06	34.0147		H_2O_2
Sodium formaldehyde sulfoxylate	0.05	154.12		CH_7NaO_5S
Di butyl phthalate	1.0	278.34		$C_{16}H_{22}O_4$
Formaline	0.1	30.031		CH_2O
Biocide	0.05	*biocide*	N/A	Biocide

Silquesit A-171 (Vinyltrimethoxysilane)	0.1	148.23		$C_5H_{12}O_3Si$
Defoamer	0.06	Defoamer	N/A	Defoamer
Total Percentage	100	N/A	N/A	N/A

Sodium formaldehyde sulfoxylate is used in redox-reaction polymerization as a reducing agent and as a release and stripping agent in the textile industry; it is also used in emulsion polymerization as a reducing agent, and it can be used in combination with all other oxidizing agents, and the recommended usage would be within the range 0.05-0.2% for post-polymerization reactions, and it can be dry stored for 12 months at 25 C.

Di butyl phthalate is used in adhesives, printing inks, and lacquers as a plasticizer, and it can be used with other phthalates for PVC compounds as a secondary plasticizer.

Formalin is used as a preservative in some types of adhesives, such as vinyl, and it may result as a by-product from some polymerization reactions. *Biocide* is known to be a chemical substance that has a controlling effect on any harmful organism by biological or chemical means.

. Additives are chemical substances added in small amounts to the main ingredient to improve some chemical or physical properties of the material being synthesized. Silquesit A-171, Vinyltrimethoxysilane, offers silanes and vinyl compounds some functionality by crosslinking organic polymers; the crosslinked Si-O-Si bond formed is highly resistant to UV light, other chemicals and exposure to moisture. It can also be added as a monomer in emulsion polymerization to form modified silane latexes, functioning as a crosslinker forming stable Si-O-Si bond linkages.

4. Discussion

The current formulation consists of the following chemical compounds, 0.4% Natrosol 250- HHBR, 44.39% water, 0.14% Sodium Bicarbonate, 2.0% Polyoxyethylene 25 octyl phenol, 2.0% Octyl phenol polyglycol ether sulfate sodium salt, 0.5%, Provichem, 2.0% Buyl acrylate, 47% Vinyl acetate, 0.1258% Potassium Persulfate,0.0234% Tertiary butyl hydrogen peroxide, 0.06% Hydrogen peroxide, 0.05% Sodium formaldehyde sulfoxylate, 1.0% Di butyl phthalate, 0.1% Formaline, 0.05% *biocide* , 0.1% Silquest A-171 and 0.06% defoamer.

The current acrylic paint formulation consists of mainly of 44.39% water as solvent, 49% pre-emulsion monomers (vinyl acetate and butyl acrylate), and 6.61% additives (Natrosol 250- HHBR, Sodium Bicarbonate, Polyoxyethylene 25 octyl phenol, Octyl phenol polyglycol ether sulfate sodium salt, Provichem, 2.0% Buyl acrylate, Potassium Persulfate, Tertiary butyl hydrogen peroxide, Hydrogen peroxide, Sodium formaldehyde sulfoxylate, Di butyl phthalate, Formaline, *biocide*, Silquest A-171, and defoamer).

The Current Percentage of each component in the current formulation provided are Pigments (Sodium bicarbonate) 0.14%, Extenders (none used), Additives 6.4629%, Solvent 44.39%, and Binders 49%.

The main pre-emulsion monomer used is of only one type of monomer, vinyl acetate and butyl acrylate are only of 2% and is used as additive in the current formulation. The binder formed from the pre-emulsion polymerization of the monomers used in the formulation would mainly be a homopolymer consisting of vinyl acetate monomers and some butyl acrylate impurities.

The amount of surfactant used as an additive is 4.5%, which is very high and causes a lower scrub resistance test for the current formulation. The addition of surfactants does not always have a positive effect on all properties. The water resistance of the coating can be decreased with surfactant addition since surfactants can be very water-soluble and will easily wash out of a coating. This problem of moisture resistance is a particularly prevalent problem for art conservation, as well as problems with adhesion, loss of optical clarity, and dirt pickup caused by polyether surfactants in contemporary acrylic emulsion used in artworks bearing acrylic coats. While the type and amount of surfactant determine what properties will be affected, other chemicals in a paint can alter the overall effect the surfactants may have on the paint. Elasticity has been found to either increase or decrease in latex paints depending on the amount of TiO_2 present.

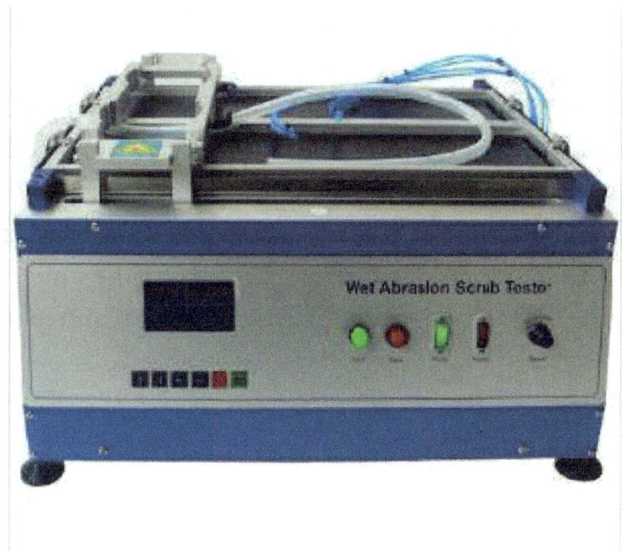

Figure1. Sheen Scrub Tester used in the scrub resistance test for the current acrylic paint formulation.

5. Conclusion

In order to improve the quality of the paint and increase its scrub resistance, missing chemical compounds from the current formulation, pigments, extender, and another type of pre-emulsion monomer, must be added to the formulation. The use of pigments and extenders in the formulation would form a firm protective layer and prevent the dissolution of the paint film from the surface, and to improve the scrub resistance test of the acrylic paint and it's recommended to avoid using copious amounts of surfactants.

The current formulation consists of pigments 0.14% in the form of Sodium Bicarbonate, compared to the literature amount 25%. Extender pigments weren't used in the current formulation compared to the literature amount 12%, surfactants 4.5% compared to the literature amount 0-1%, and binders 49% compared to the literature amount 14%. The current paint scrub resistance test shows the paint operates with 60% effectiveness compared to the reference due to missing 40% of the main ingredients that make good acrylic paint.

Copious amounts surfactants 4.5% (2.0% Octyl phenol polyglycol ether sulfate sodium salt, 2.0% Octyl phenol Polyglycol Ether Sulphate Sodium Salt, and 0.5% Provichem) have been causing a low scrub resistance test of current formulation compared to the reference. The large amounts of surfactants are used as an additive, 4.5%, has a negative effect on the formulation, causing a low scrub resistance test result compared to the reference.

Chapter 3
Increase Product Quality for a Car-wash Shampoo Concentrate

1. Overview

The market for vehicle cleaning products in Western Europe approached a value of $400 million dollars in 2007. Domestic and industrial automated cleaning of vehicles can include four main steps, pre-wash, main wash, rinse, and drying, besides the manual cleaning of domestic vehicles. This chapter presents the methods used to improve the properties of the current car wash shampoo concentrate formulation and increase its quality. The chapter also presents some new chemical compounds and additives that would increase the quality of the car wash shampoo concentrate and methods used to increase the concentration of the current formulation for its use and distribution.

2. Introduction

Saka International Group, a leading manufacturer of home care business, laundry, and car care products by Schnnell, ON, Canada, are looking to improve the chemical properties of their Car wash shampoo. A liter of carwash shampoo concentrate dilutes only to 200L using the current formulation. The formulation consists of the following chemical compounds: 5% Linear Alkyl Benzene Sulphonic Acid (LABSA), 20% Sodium Lauryl Ether Sulphate (SLES), 5% Sodium Lauryl Sulphate (SLS), 1% Sodium Hydroxide, 2% Betaine, 61.5% Water, 3% Sodium triphosphate, 2% Glycerol, and 0.5% Propylene Glycol.

Car maintenance products are classified into the interior and exterior car care products. Interior car care products are deodorants, grease cleaners, vinyl and plastic cleaners and polishes, and interior wins screen cleaners, carpet shampoos, and leather polishes. Interior car care products include tyre dressings and cleaners, pre-soak detergents, car polishes, wash and wax formulations, windscreen cleaners, water repellents and drying aids, and wheel rim cleaners. The climate and the season of the year affect the nature of the soiling of the vehicle and the ease of its removal.

The bodywork of the automobile consists of multiple coatings; each coating provides a variety of functions. Figure 1 shows the coating layers in the bodywork of modern cars. The paintwork of the vehicle is the external surface to be cleaned.

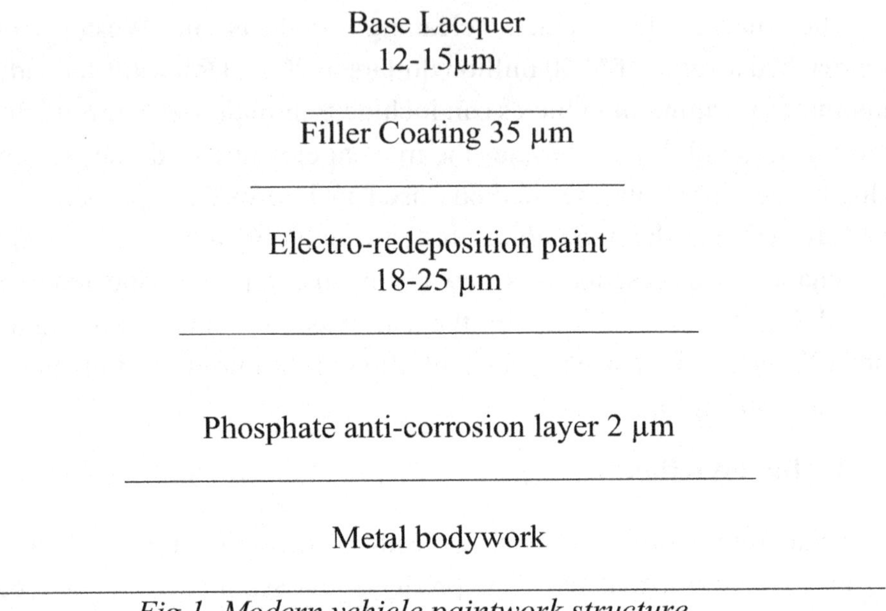

Fig 1. Modern vehicle paintwork structure.

The base coat is usually water-based polymeric binders, fillers, and pigments. The inner coating, the electro-deposition paint, and the phosphate-based anti-corrosion layer provide protection to the metal surface. On top of the protective coatings is the filler layer; it must have an excellent adhesion property to both the top coat and the base coat. The finishing lacquer must have good impact strength, retain gloss, and it must be waterproof. Domestic and industrial automated cleaning of vehicles can be divided into four main steps, pre-wash, main wash, rinse, and drying, besides the manual cleaning of domestic vehicles. Pre-wash includes cold degreaser, microemulsion, and foam wash. The main wash includes shampoo and microemulsion. Rinse includes hot/ cold wax and rinse aid.

Car shampoos can be either in liquid or powder form. Liquid car shampoos are a combination of binders, surfactants, and liquids dissolved in water as the main solvent. These products are easy to rinse off, high foaming, biodegradable, made to cut through grease on the bodywork, and they don't damage any part of the vehicle surface, including the paintwork. Economy car shampoos do not contain builders. Powder car shampoos are made of a mixture of builders (carbonates, phosphates, or metasilicates) and surfactants (fatty alcohol ethoxylates or dodecylbenzene sulphonates) absorbed into the powder.

The main anionic detergent can be either Alkylbenzene sulphonates and/or sodium lauryl ether sulfates. Sodium lauryl ether sulfates is incorporated into the formulation used when the denser, richer foam is required. Fatty acid alkanolamides, amine oxide, or betaine are used in the formulation for viscosity and to stabilize the foam produced. For greater foam stability and viscosity, amides are added to the formulation. To increase the quantity of the foam produced, betaines and amine oxides are used. Glycerol ether is used to ease grease removal. Secondary surfactants are used for viscosity and foam modifications and are also used to enhance spot removal and improve detergency. Binders such as phosphates (0.5-2.5%) are added to improve detergency.

Table 1. Traditional anionic car wash shampoo formulation.

Chemical Compound	% by weight used the formulation
Sodium carbonate (binders)	2
Sodium metasilicate pentahydrate (binders)	3
Sodium citrate (water softner)	2
Glycerol ether (solvent)	4

Linear alkyl benzene sulphonate (30%) detergent	27
Sodium lauryl ether sulphate (28%) Detergent	10
Coconut diethanolamide (foam producer)	3
Water	49
Preservatives/ dyes	Q.S.

Low hydrophilic-lipophilic balance (HLB) fatty alcohol ethoxylate/ hydrotropic system replaced the traditional anionic surfactant-based car shampoo as they afford more effective cleaning performance, decreasing and they have low foam profile.

Table 2. HLB fatty alcohol ethoxylate/ hydrotropic car wash shampoo formulation.

Chemical Compound	% by weight used the formulation
Fatty alcohol ethoxylate(low HLB)	5
Hydrophobe (alkyl glycoside or quaternary fatty amine ethoxylate)	5-10
TKPP (Tetrapotassium Pyrophosphate) Detergent builder	6
Sodium metasilicate	4
Balance water	

Sodium citrate is known to be a water softener and a PH adjuster. It's an ingredient in most common liquid detergents. It's also used in some food products to adjust their acidity. It's used in ice cream, gelatin desserts, candy, and jelly. It's also used in some pharmaceutical and personal care products, such as sunscreens, facial moisturizers, makeup, baby wipes, soaps, shampoos, and conditioners.

Coconut diethanolamide is used as an emulsifying and foaming agent in personal care products and in cosmetics. It's extracted from coconut oil. It's also used in hydraulic fluids and industrial cooling lubricants.

3. Experimental

The chemical composition of the ingredients used in the making of the car-wash shampoo concentrate is presented in table 1.

Linear alkyl benzene sulphonic acid is used as a mercerizing and washing agent in the textile industry. It's used as an emulsifier and wetting agent in small quantities with surfactants as it increases the surface area of distempers. ..... complete Because of its good performance and low-cost linear alkylbenzene sulfonic acid is the largest volume synthetic anionic detergent. As all surfactants, linear alkylbenzene sulfonic acid has both hydrophilic and hydrophobic groups. Other examples of commercial anionic surfactants are alkyl sulfates and Alpha-olefin sulfonates. These compounds are produced by sulfonation, and they are non-volatile compounds. Linear alkylbenzene sulfonic acid consists of a phenyl isomer of 5 to 2 position substituents, different alkyl chain lengths consisting of ten to fourteen carbon atoms (C10-C14), and an aromatic ring sulfonated at the para position attached to the linear alkyl chain at any position except position 1, the 1-phenyl position. The chemical and physical properties of linear alkylbenzene sulfonic acid differ based on the length of the alkyl chain, giving rise to different formulations and different usage in various applications.

Sodium Lauryl Ether Sulphate is a surfactant, and it's an anionic detergent. It's a very effective foaming agent, inexpensive, and used in shampoos, toothpastes, and soaps.

Sodium Lauryl Sulphate is used in hygiene, cleaning, food, and pharmaceutical products, and it's an anionic surfactant. It is widely used as a food additive in the food industry and as an emulsifier and an ionic solubilizer in the pharmaceutical industry.

Sodium Hydroxide Sulphate is a strong base, hygroscopic solid, soluble in water, and can cause severe burns. It's used in various industries; it's used in the manufacture of drain cleaners, soaps, detergents, textiles, drinking water, pulp, and paper. It's used as a paint stripper, cleaning agent, relaxer, and in food preparation.

Betaine is a chemical compound occurring in plants, and it's an amino acid; it's a white solid at room temperature . It is present in living cells, and it's a methylated nitrogen compound. It's used in the pharmaceutical industry in the preparation of shampoo and soap as it's a nonionic surfactant. Due to its high surface activity, its derivatives are used as efficient cleansing agents. It's also used as viscosity modifiers, foam stabilizers, and detoxifiers. Betaine esters can be used in antiperspirants as it processes anti-microbial activity.

Water is an inorganic compound that is odorless, tasteless, transparent liquid at room temperature and acts as a solvent. . It's present in 70% of the earth's surface as seas and oceans. In the world economy, 70% of water is used in agriculture.

Sodium triphosphate is used as a component of industrial and domestic products on a large scale.

It's used as a builder and a water softener in commercial detergents. Detergents are deactivated in water containing high concentrations of Mg^{2+} and Ca^{2+}, hard water. It's a chelating agent as it binds tightly to bications and prevents them from interfering with sulfonate detergent. It's used as an emulsifier to retain moisture, and it's also used as a preservative in the food industry. It's also used as an anti-cracking agent, flame retardant, anti-corrosion pigment, synthetic tanning agent, and masking agent in the leather industry.

Glycerol is an odorless and non-toxic colorless viscous liquid at room temperature. Glycerol is hygroscopic in nature and is miscible in water. It's used as a humectant in pharmaceutical formulations as it improves the ability of the skin to absorb water, and it's used in the food industry as a sweetener.

Propylene Glycol is a colorless viscous liquid at room temperature and has a faintly sweet taste. It's miscible in a wide variety of solvents including chloroform, acetone, and water. It's a non-irritating substance with low volatility. It is used in various industries, including food and drug, anti-freezes, polymers, and electronic cigarettes. It's used as a humectant in hand sanitizers to prevent skin drying. It's used in coffee-based drinks, ice creams, whipped dairy products, soda, and liquid sweeteners. Polypropylene glycol alginate gives rise to a greater increase in form stability equal to the amount of neutral polysaccharides.

Table 2, lists the Chemical name, molecular formula and molecular weight of chemical compounds used in the current formulation.

Chemical Name	Chemical structure	Molecular Formula	Molecular Weight (g/mol)	Amount used in the current formulation (%)
Linear Alkyl Benzene Sulphonic Acid (LABSA)		$CH_3(CH_2)_{11}C_6H_4SO_3H$	326.49	5
Sodium Lauryl Ether		$CH_3(CH_2)_{11}(OCH_2CH_2)_nOSO_3Na$	Variable; typically around	20

			421 g/mol	
Sulphate (SLES)				
Sodium Lauryl Sulphate (SLS)	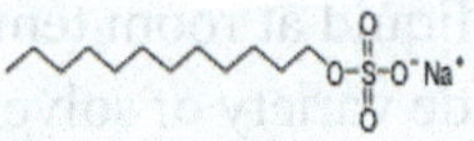	$C_{12}H_{25}NaSO_4$	288.372	5
Sodium Hydroxide.		NaOH	39.997	1
Betaine.	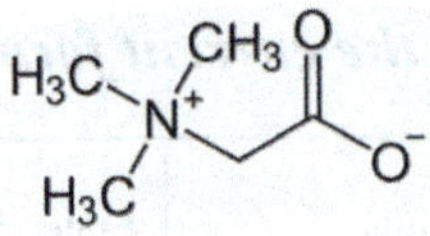	$C_5H_{11}NO_2$	117.146	2
Water (Dihydrogen Monoxide)		H_2O	18.01	61.5
Sodium triphosphate	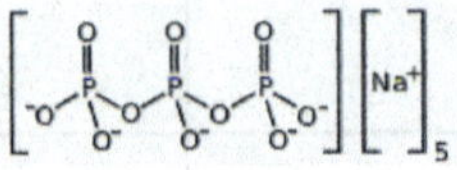	$Na_5P_3O_{10}$	367.864	3

| Glycerol | | $C_3H_8O_3$ | | 92.094 | 2 |
| Propylene Glycol | | $C_3H_8O_2$ | | 76.095 | 0.5 |

The current formulation consists of 1% builder in the form of sodium hydroxide, 3% water softener in the form of sodium triphosphate, 32% surfactants in the form of 5% Linear Alkyl Benzene Sulphonic Acid, 20% Sodium Lauryl Ether Sulphate, and 5% Sodium Lauryl Sulphate, and 2.5% solvents in forms of 0.5% Propylene Glycol and 2% Glycerol.

Sodium and potassium hydroxides are known to be used in wheel rim cleaner formulations as the alloy wheel picks up dirt and grease from the road, and they are prone to dirt from the abrasive wear of brake shoes. The amount of sodium and potassium hydroxides used in the wheel rim cleaner formulation is 0-15%.3. It isn't known to be used in carwash formulations. Other builders are known to be used in car wash formulations to soothe out slight imperfections and to remove road grease and stubborn tar from the bodywork of the vehicle. These builders would be calcium carbonate, silicones, and lamella aluminum silicates. To increase the quality of the car wash shampoo formulation, silicone derivative builders are added to the formulation. Silicone derivative builders contribute to the ease of application of the products, the gloss, and it's water repellency property.

The amount of surfactant used in the current car wash shampoo formulation concentrate is 32% (5% Linear Alkyl Benzene Sulphonic Acid, 20% Sodium Lauryl Ether Sulphate, 5% Sodium Lauryl Sulphate, and 2% Betaine). To increase the quality of the car wash shampoo, the amount of surfactants in the formulation must be increased to 37%.

The solvents used in the current formulation: 0.5% propylene glycol and 2.0% glycerol, are both known to be used as humectants in pharmaceutical and personal care product formulations to improve the ability of the skin to absorb water. These solvents aren't known to be used in car wash shampoo formulations. Other solvents used in car wash shampoo formulations (4% glycol ether) to aid in the removal of surface dirt and to act as a carrier for silicates and components of the shampoo. Solvents that are commonly used in car wash shampoo formulations are glycol ethers such as di propylene alcohol mono methyl ether, used as a solvent to dissolve grease.

The choice of solvent is critical to avoid the stress and cracking of plastics and to avoid damage to the painted surface. It isn't recommended the use glycerol or propylene glycol in the formulation, and it's recommended the use of glycol ether. Also, the amount of solvents used in the current formulation is low, 2.5%, compared to the amount used, reference 4%. The amount of solvents used should be increased to 4% to increase the quality of the car wash shampoo formulation.

For the current formulation to dilute to a greater volume of 500L, and to also maintain the quality of the car wash shampoo concentrate, the amount of surfactants, builders, foaming agents, and solvents must be increased relative to the amount of the main solvent used to dissolve it's components. The current formulation dilutes to 200L; in order for the formulation to dilute to double this volume, the amount of the main solvent used to dissolve it's components must be decreased to ½, and the amount of surfactants/ builders/ foaming agents and solvents must be increased by 1 ½. Taking the traditional anionic car wash shampoo formulation as an example, the percentage of each chemical compound in the formulations would be 3% Sodium carbonate, 4.5% Sodium metasilicate pentahydrate, 3% Sodium citrate, 6% Glycerol ether, 40.0% Linear alkyl benzene sulphonate, 15% Sodium lauryl ether sulphate, 4.5% Coconut diethanolamide, and 24.0% water as the dissolving solvent.

4. Conclusion

To increase the quality of the car wash shampoo- concentrate 2% builders in the form of sodium carbonate, 3% builders in the form of silicates, and 3% foaming and emulsifying agents in the form of Coconut diethanolamide, should be included in the formulation. It's also recommended the use of another different solvent to dissolve grease and improve the quality of the car wash shampoo concentrate, 4% glycol ether.

Foaming and emulsifying agent are used to dissolve solvents, builders, and silicates and facilitate the formation of foams. Builders used in car wash shampoo formulations remove road grease and stubborn tar from the bodywork of the vehicle, smooth out minor surface scratches and slight imperfections, contribute to the gloss, the water repellency, increase the durability of the overseal, and polish the paintwork. The choice of builders and solvents is critical to avoid the stress-cracking of plastics and to avoid damage to the painted surface. Calcium carbonates, silicones, lamella aluminum silicates, and poly dimethyl siloxane are known to be used in car wash shampoo formulations as builders.

It's recommended to avoid the use of sodium hydroxide as a builder or the use of glycerol/ propylene glycol as solvents in the car wash shampoo formulation, and the amount of surfactants must be increased to at least 37%.

Sodium hydroxide is known to be used in wheel rim cleaner formulation as the alloy wheel pick up dirt and grease from the road, not in car wash shampoos; glycerol and propylene glycol are known to be used as humectants in lotions and personal care products. Sodium hydroxide, glycerol, and propylene glycol aren't known to be used in car wash shampoo formulations.

In order to increase the concentration of the car-wash shampoo concentrate, and make a liter of concentrate dilute to 500L instead of 200L, the amount of builders, solvents, and surfactants must be increased to 1 ½, and the amount of water must be decreased to ½.

Chapter 4
Modification to An Acrylic Paint Formulation

1. Overview

Paints are used to protect and prolong the life of natural and synthetic materials as it acts as a barrier against environmental conditions. Paints contain extenders, solvents, pigments, binders, and some additives. The contents of acrylic white matt emulsion paint are known to be 25% Pigments, 12% Extender pigments, 5% Additives, 44% Solvents, and 14% Binders.

Binders are matrices and are used to hold the pigment in place. Extenders have larger pigment particles to improve adhesion, strengthen the film, and save the binder. Pigments are used to give color and opacity to the paint. A solvent can be either an organic solvent or water and is used as a thinner to dissolve paint components and make them uniform. Additives are commonly used to improve the properties of the paint.

This chapter briefly shows how to manipulate the chemical and physical properties of chemical compounds making acrylic matt paint improve its quality. It presents the chemical compounds that would potentially lower the scrub resistance test. It also presents the additives and the chemical compounds that would potentially increase the quality and increase the scrub resistance test for acrylic matt paint.

2. Introduction

A polymer production unit in Saudi Arabia, PSCF, is looking to improve its current formulation of the current acrylic matt paint formulation as the current formulation shows a low scrub resistance test compared to the reference. The current paint scrub-resistance test shows the paint operates with 60% effectiveness compared to the reference.

The current formulation consisted of 47% Vinyl acetate, 0.1258% Potassium Persulfate, 0.0234% Tertiary butyl hydrogen peroxide, 0.06%

Hydrogen peroxide, 0.05% Sodium formaldehyde sulfoxylate, 1.0% Di butyl phthalate, 0.1% Formaline, 0.05% biocide, 0.1% Silquest A-171 and 0.06% defoamer , 0.4% Natrosol 250- HHBR, 44.39% water,0.14% Sodium Bicarbonate, 2.0% Polyoxyethylene 25 octyl phenol, 2.0% Octyl phenol polyglycol ether sulfate sodium salt, 0.5%, Provichem, and 2.0% Buyl acrylate. Chemical names, molecular formulas, and molecular weights of all chemical compounds used in making the current acrylic paint formulation are listed in table 1.

An acrylic white matt emulsion paint is known to consist of 25% Pigments, 44% Solvents, 12% Extender pigments, 5% Additives, and 14% Binders, Figure 1

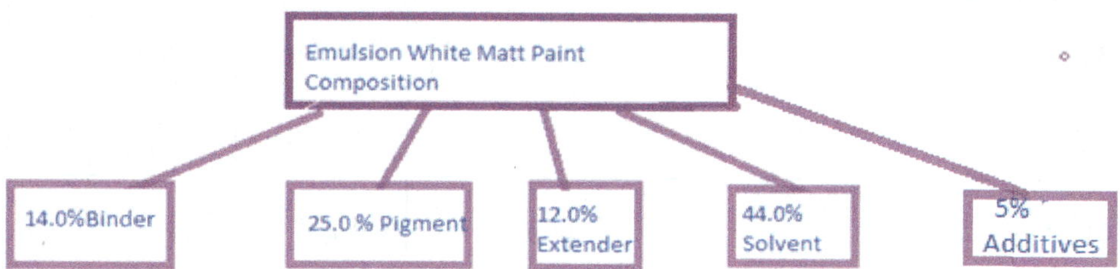

Figure 1. Chemical Composition for a matt white Paint

Table 1. lists the chemicals, amounts in percentage, and the role of chemical compounds used in the current acrylic paint formulation.

Chemical Name	Percent (%) used in the current formulation	Molecular Formula	Role of Each Chemical Compound in Current Formulation
Water	44.09	H2O	Solvent
Natrosol 250- HHBR	0.4	$C_{36}H_{70}O_{19}$	Nonionic water thickener
Sodium Bicarbonate	0.14	$NaHCO_3$	Pigment

Polyoxyethylene (25) octyl phenyl ether	2.0	(C2-H4-O)mult-C14H22-O	Surfactant and buffer
Octyl phenol polyglycol ether sulfate sodium salt	2.0	Octyl phenol polyglycol ether sulfate sodium salt	Surfactant and buffer
Provichem (*Sodium vinylsulfonate*)	0.5	$C_2H_3NaO_3S$	Surfactant
Buyl acrylate	2.0	$C_7H_{12}O_2$	Monomer in homo or copolymerizati on (Binder)
Vinyl acetate	47.0	$C_4H_6O_2$	Monomer in homo or copolymerizati on (Binder)
Potassium Persulfate	0.1258		Initiator
Tertiary butyl hgrogen peroxide	0.0234	$C_4H_{10}O_2$	Oxidizing catalyst
Hydrogen peroxide	0.06	H_2O_2	Initiator
Sodium formaldehyde sulfoxylate	0.05	CH_7NaO_5S	Reducing agent.
Di butyl phthalate	1.0	$C_{16}H_{22}O_4$	Secondary plasticizer
Formaline	0.1	CH_2O	Preservative
Biocide	0.05	Biocide	Biocide
Silquesit A-171 (Vinyltrimethoxysilane)	0.1	$C_5H_{12}O_3Si$	Crosslinking polymer

Defoamer	0.06	Defoamer	Defoamer
Total Percentage	100	N/A	N/A

3. Detailed Analysis

The current acrylic paint formulation consisted of mainly of 44.39% water as a solvent, 49% pre-emulsion monomers (vinyl acetate and butyl acrylate), and 6.61% additives (Natrosol 250-HHBR, Sodium Bicarbonate, Polyoxyethylene 25 octyl phenol, Octyl phenol polyglycol ether sulfate sodium salt, Provichem, 2.0% Buyl acrylate, Potassium Persulfate, Tertiary butyl hydrogen peroxide, Hydrogen peroxide, Sodium formaldehyde sulfoxylate, Di butyl phthalate, Formaline, biocide, Silquest A-171, and defoamer).

4. Conclusion

The current formulation consists of pigments 0.14% in the form of Sodium Bicarbonate, compared to the literature amount 25%. Extender pigments weren't used in the current formulation compared to the literature amount 12%, surfactants 4.5% compared to the literature amount 0-1%, and binders 49% compared to the literature amount 14%. The copious amounts surfactants are causing a low scrub resistance test of the current formulation compared to the reference.

In order to increase the quality of the acrylic paint, pigments, extenders, and another type of pre-emulsion monomer should be included in the formulation to form a firm protective layer and prevent the dissolution of the paint film from the surface and improve the scrub resistance test of the acrylic paint. It's recommended to avoid using copious amounts of surfactants, 4.5% (2.0% Octyl phenol polyglycol ether sulfate sodium salt, 2.0% Octyl phenol Polyglycol Ether Sulphate Sodium Salt, and 0.5% Provichem).

Chapter 5
Modification to a Chemical Formulation for Optimal Performance

1. Overview

In a chemical reaction, the reactant substance changes to a product substance, and the product substance will have different physical and chemical properties than the reactant substance. All chemical reactions involve a detectable change: color change, bubbling, heat evolution, heat absorption, light emission, or formation of a precipitate. Reactions are classified into three main reactions, precipitation reactions, acid-base reactions, and oxidation-reduction reactions. The counting unit for a number of atoms, ions, or molecules in a laboratory seize sample is mole, abbreviated mol. A limiting reactant is a reactant that is completely consumed in a chemical reaction; it limits and determines the amount of product formed; the other reactants are called excess reactants. The quantities of product formed and reactant consumed are restricted by the amount of limiting reactant. This Chapter presents the methods used to increase the quality of a bubble-forming product, increasing the amount of bubbling per table by using the difference in chemical and physical properties of the chemical compounds included in the ingredient. The chapter also presents chemical compounds in the ingredient that would be harmful to the quality of the product, and it also presents new additives that would increase the quality of the product.

2. Introduction

Some of the companies in the US are looking to produce kid-safe foaming tablets from household chemicals that would produce bubbles when placed in the snow. The ingredient included the following chemical compounds; corn starch (pure starch), red beet powder (Betanin), salt (Sodium Chloride), cream tartar (Potassium Bitartrate), baking soda (Sodium Bicarbonate), Vitamin C (citric acid), drops of water (H_2O). The amount of bubbling was minimal; also, the tablet didn't hold together and was grainy and fragile. Since bubbling forms when the fragile tablet is placed in the snow, that is evidence of a chemical reaction.

In a chemical reaction, the reactant substance change to a product substance, and the product substance will have different physical and chemical properties than the reactant substance. All chemical reactions involve a detectable change.

A chemical equation is used to describe a chemical reaction. A Chemical reaction a be represented by a molecular equation. Reactant species are always listed on the left side of a chemical equation and the product species are listed on the right side of the chemical equation and they are separated by an arrow. A chemical reaction can also be presented by an ionic equation or net ionic equation.

Reactants → Products.

Reactions are classified into three main reactions, precipitation reactions, Acid-base reactions, and oxidation-reduction reactions. Precipitation reactions would involve the formation of insoluble ionic compound forms from mixing a solution of two ionic substances. Acid-base reactions involve the reaction of an acid with a base and a transfer of a proton between reactants are involved. Oxidation-reduction reactions (redox reactions) involve the transfer of an electron between reactant species.

The reaction that takes place when placing the created tablet in water/ice is an acid-base reaction. Acid-Base reactions are widely known for their ability to produce bubbling if either reactant is a weak acid or weak base that would produce a gas product. In an acid-base reaction, acid and base properties neutralize, where the anion of the acid and the cation of the base combine to form the salt, and the hydroxide anion, OH^-, combines with the hydrogen cation, H^+, to form water.

The counting unit for a number of atoms, ions, or molecules in a laboratory seize sample is mole, abbreviated mol. A mole of samples of different substances has different masses. For example, 1 mole of 24Mg and 1 mole of 12C. A single 24Mg atom is twice as massive and has a mass of 24 a.m.u, as a single 12C atom has a mass of 12 a.m.u. The formula weight of any substance is numerically equal to the molar mass of that substance. For example, for table salt with the chemical formula NaCl, the formula weight is 58.5 a.m.u, and its molar mass is 58.5 g/mol. The molar mass of a substance in g/ mol can be used as a conversion factor to convert the mass of a substance into moles or moles of a substance into grams.

The relative number of molecules in a reaction is represented by coefficients in a balanced chemical equation. The relative number of moles and the relative number of molecules in a reaction is indicated by the coefficients in a balanced chemical equation.

A limiting reactant is a reactant that is completely consumed in a chemical reaction. It limits and determines the amount of product formed. The other reactants are called excess reactants. The quantities of product formed and reactant consumed are restricted by the amount of limiting reactant. If one reactant, the limiting reactant, is consumed, the reaction stops, and the amount of limiting reactant becomes zero at the end of the reaction.

Carboxylic acids are the largest category of weak organic acids that contain carboxylic groups (-COOH). The generic formula for carboxylic acids is R-COOH, where R= alkyl group.

The acidic behavior of carboxylic acids is due to the oxygen atom bonded to the carboxyl group carbon drawing electron density from the hydroxyl bond. It helps stabilize the conjugate base and increases the polarity of the -OH bond. The carboxylate anion, the conjugate base of acid, has resonance, which spreads the negative charge over several atoms and contributes to the stability of the anion.

Alcohols are compounds that contain one or more hydroxyl or alcohol groups (-OH). The -OH bonds are polar; bonds and alcohols are more soluble in polar solvents than hydrocarbons, and OH groups can form hydrogen bonds. Examples of alcohols are 1,2 ethanediol and glycerol

In Industry, occasional modification to the formulation of a product is necessary to improve its quality and performance. In 2021 Chegeni and group developed a new formulation for air revitalization tablets using the Taguchi statistical method in closed atmospheres. The air revitalization tablets are used in closed spaces such as submarines, underground mines, shelters, and spacecrafts to absorb carbon dioxide, decrease its concentration, and produce fresh oxygen necessary for life in a closed space. The initial formulation consisted of various amounts of additives: MnO_2 (catalyst), LiOH (CO_2 auxiliary absorbent), $Cu_2Cl(OH)_3$, and $CaSO_4$. It has been found that for optimal performance, the greatest amount of oxygen is produced, and the maximum amount of carbon dioxide adsorbed occurs when using a certain amount of each chemical compound in the formulation. This optimal formulation consisted of 3 wt% lithium hydroxide, 3 wt% copper oxychloride, 3 wt% calcium sulfate, and 5 wt% manganese dioxide.

In medicine, modification to the method of delivery of the drug to the organ is necessary to improve its performance. In 2020, Chen and the group developed an optimal method to deliver the drug formulation to the pancreas; they used a bubble bursting-mediated oral drug delivery system that enables concurrent delivery of hydrophilic and lipophilic chemotherapeutics for treating pancreatic tumors in rats. The optimal formulation consisted of PTX (1 mg) that had been pre-dissolved in capric acid (18 mg), GEM (1 mg), sodium bicarbonate (5 mg), and citric acid (0, 2, 4, 6, or 8 mg) was exposed to a simulated intestinal fluid (SIF; 3 mL, pH 6.4) that contained bile extract (5 mg/mL). The oral drug delivery system initiates an effervescent reaction to form gas-bubble carriers were proposed; these carriers deliver lipophilic paclitaxel (PTX) and hydrophilic gemcitabine (GEM) in the small intestine. The bursting of the bubbles promotes the absorption of the drugs in the intestines. The study revealed that the orally delivered formulation has no toxic side effects that are associated with the i.v. injected formulation. It resulted in an increase in the bioavailability of PTX, and the oral formulation had a greater impact than the i.v. formulation on tumor-specific stromal depletion, resulting in greater inhibition of tumor growth with no evidence of metastatic spread. This unique approach of oral chemotherapy has the potential for use on outpatients, greatly improving the quality of their lives and enhancing therapeutic efficacy.

3. Experimental

The chemical composition of the ingredients used in the making of the bubble-producing tablets is presented in table 1.

Corn starch is pure starch and is known to be used as a liquid thickener. Red beet powder is used as a dye that colors the tablet. Table salt is sodium chloride with the formula $NaCl$ and molar mass of and this salt has a neutral PH, PH = 7. Cream tartar is potassium bitartrate with formula $KC_4H_5O_6$ and molar mass: 188.177 g/mol. Baking soda is a weak base with a formula of $NaHCO_3$ and a molar mass: 84.007 g/mol. Citric acid is a triprotic acid, a weak organic acid that has the formula of $C_6H_8O_7$ and a molar mass: 192.124 g/mol.

Table 1 lists the Chemical name, molecular formula, and molecular weight of household chemicals in the ingredients

Household Chemical and Chemical Name	Chemical structure	Molecular Formula	Molecular Weight (g/mol)
Corn Starch (Pure Starch)	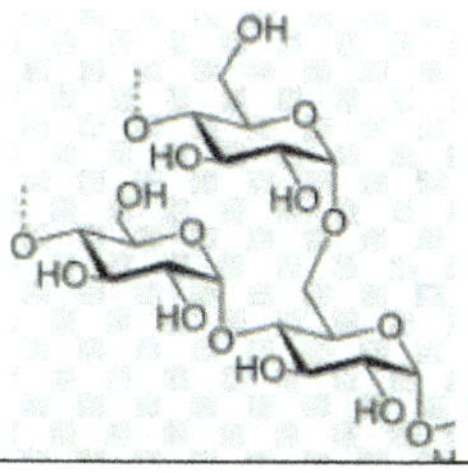	$C_{27}H_{48}O_{20}$	692.66
Table Salt (Sodium Chloride)	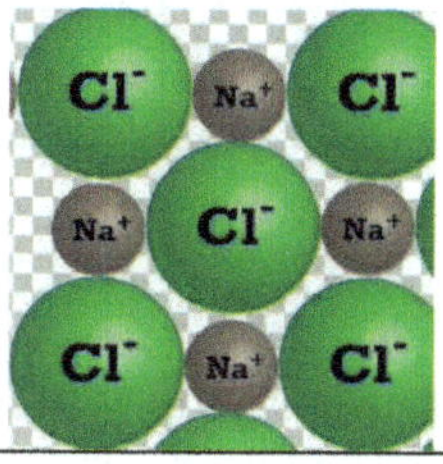	NaCl	58.44
Cream Tartar (Potassium Bitartrate)	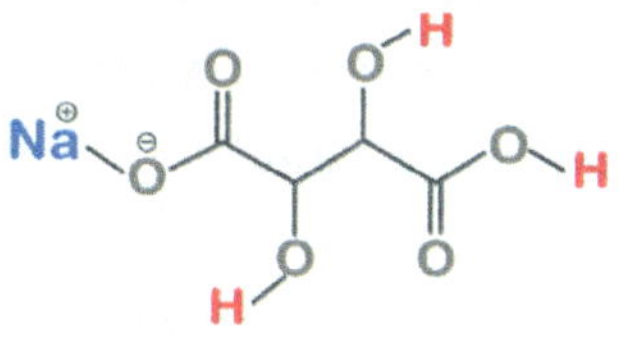	$KC_4H_5O_6$	188.177
Baking Soda (Sodium Bicarbonate)	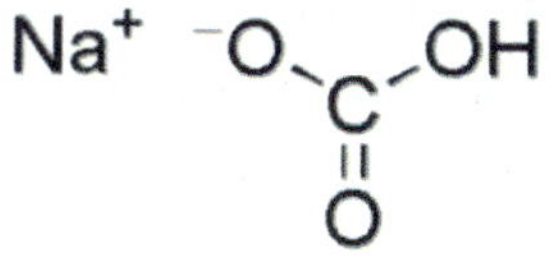	$NaHCO_3$	84.007
Citric Acid (vitamin C)	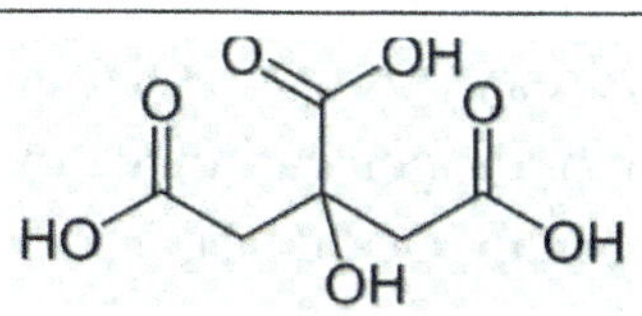	$C_6H_8O_7$	192.124
Water (Dihydrogen Monoxide)	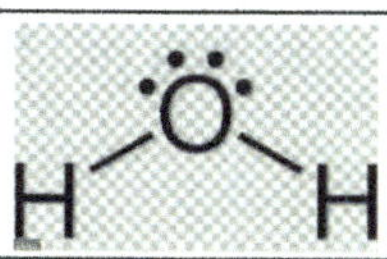	H_2O	18.01

Red Beet (Betanin)	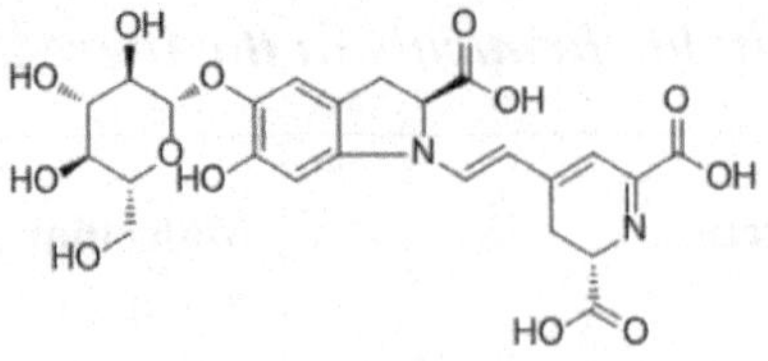	$C_{24}H_{26}N_2O_{13}$	550.47
Glycerol	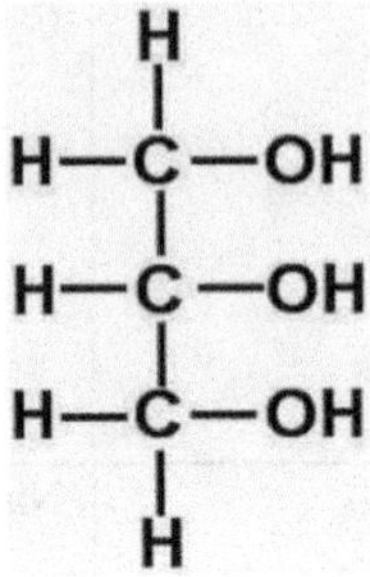	$C_3H_8O_3$	92.094

The bubbling while creating the tablet using water as a binding agent is due to the reaction between weak acid (citric acid) and potassium bitartrate with the weak base (baking soda) to form a salt, water, and carbon dioxide, causing loss of bubbles while making the tablet.

In order to get the maximum amount of yield, the maximum amount of CO2 per tablet, intern the greatest amount of bubbling, the number of moles of Cirtic acid to sodium hydrogen carbonate must be 1:3, and the number of moles of sodium bitartrate to sodium hydrogen carbonate must be 1:1 according to the balanced chemical equations.

Vitamin C, citric acid is a triprotic acid containing three acidic groups and one hydroxyl group. The hydroxyl groups (-OH) are alcoholic groups, and they don't react with Sodium hydrogen carbonate or baking soda, while the acidic groups, carboxylic group (-COOH), can react with Sodium hydrogen carbonate to form a salt, carbon dioxide, and water. The bubbling is due to the formation of carbon dioxide as a product.

The molar amount of citric acid and sodium bicarbonate reactants must be 1 : 3 in order to produce the maximum amount of bubbling in the formula per tablet:

Citric acid + 3 NaHCO$_3$ → (sodium citrate) + 3H$_2$O + 3CO$_2$

Using 1 g of vitamin C with formula $C_6H_8O_7$ in the ingredients, the amount needed for a reaction to go to completion and produce the maximum amount of carbon dioxide, and hence the maximum amount of bubbling would be 1.3 grams of baking soda as follows:

1 g /192.124 g/mol * 3 NaHCO$_3$ /1 $C_6H_8O_7$ * 84.007 g/mol= 1.312 g of NaHCO$_3$

Cream tartar, potassium bitartrate is a monoprotic acid containing an acidic group and two hydroxyl groups. The hydroxyl groups (-OH) are alcoholic groups, and they don't react with Sodium hydrogen carbonate or baking soda, while the acidic group, carboxylic group (-COOH), can react with Sodium hydrogen carbonate to form a salt, carbon dioxide, and water. The bubbling is due to the formation of carbon dioxide as a product.

Using 1g of cream tartar, potassium bitartrate with formula $KC_4H_5O_6$ in the ingredients, the amount needed for a reaction to go to completion and produce the maximum amount of carbon dioxide, and hence the maximum amount of bubbling would be grams of baking soda as follows:

(potassium bitartrate) + NaHCO$_3$ → (sodium salt) + H2O

1g/ 188.17g/mol * 1 NaHCO3/1 KC₄H₅O₆ * 84.007 g/mol= 0.4464 g of NaHCO₃.

The number of grams of sodium hydrogen carbonate needed to completely neutralize 1 g of vitamin C and 1 g of sodium bitartrate must be 1.3117g+ 0.4464g = 1.76g.

Water as a binding agent causes bubble loss while creating the tablet, and glycerol would act as a better binding agent since sugars would cause bubbling to persist, and the compounds in the ingredients are inert in glycerol. Sodium chloride reduces bubble formation, but the amount of base used was insufficient to neutralize both citric acid and Potassium bitartrate in the formula.

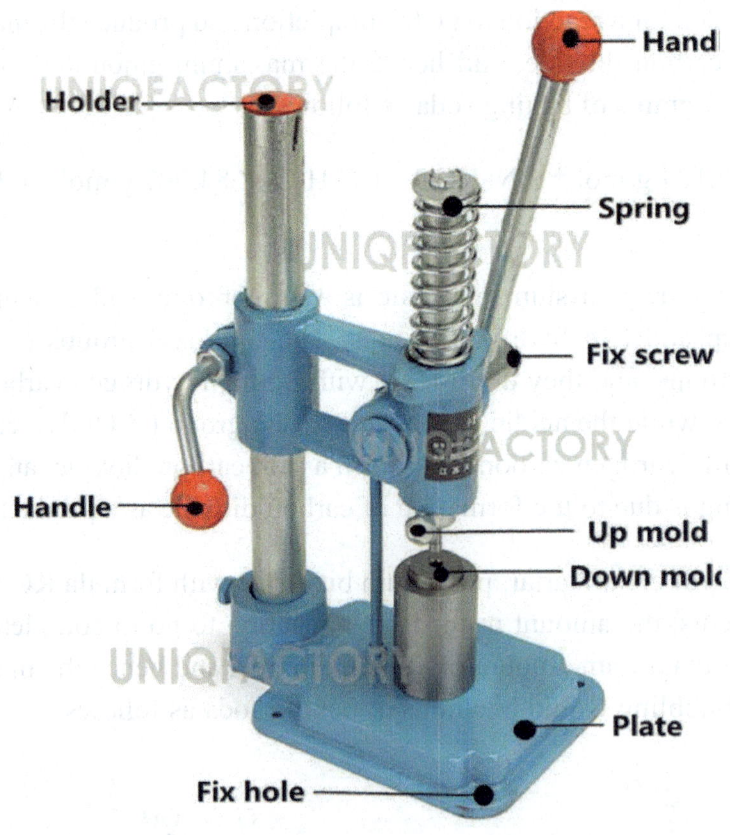

Figure1. Manual Tablet Press Machine

4. Conclusion

For optimal performance, the molar amounts of reactants were calculated to produce the greatest amount of bubbling per tablet. The bubbling power of the tablets in water is due to the reaction between vitamin C and cream tartar with baking soda. Using 1 g of Vitamin C and 1g of Sodium bitartrate, 1.76g of Sodium Hydrogen carbonate is required to be used as an ingredient to get the maximum amount of yield, the maximum amount of CO_2 in the product, in turn, the maximum amount of bubbling per tablet in the formula.

Sugars can be used to increase the quality of the product. It is recommended to use a greater number of moles of starch for the tablet to be thicker on snow, and glycerol acts as an ideal chemical binding agent for the ingredient.

Sugars, including glycerol, sugar alcohol, and sucrose, are good additives to cause the persistence of the bubbling forming ability of the tablet created, and it is not recommended to use sodium chloride in the formulation as it would prevent forming bubbles.

For the tablet to look smoother and for equal distribution of each of the compounds in the ingredient per table, the ingredients must be grinded to powder before adding the binding agent glycerol.

The tablets have been created using the correct molar amount for each chemical in the ingredients, and the amount of bubbling has increased significantly; the amount of bubbles formed has doubled using the correct formula.

Chapter 6
Thermodynamic Study and Oxidation Products for the Reaction of Methyl Ethyl Sulfide with Oxygen at Standard Conditions

1. Overview

Methyl ethyl sulfides are used in organic synthesis and scientific research. A significant amount of ethyl methyl sulfide is emitted into the atmosphere due to biological activity in the sea, and the concentration of ethyl methyl sulfide is the result of anthropogenic and human activity. Methyl ethyl sulfides have a negative effect on the atmosphere as it produces sulfuric oxides in the environment. The study of methyl ethyl sulfide oxidation products is of value to knowing the fate of its oxidation products in the atmosphere.

This chapter presents the methods used in thermodynamically identifying the energetically favored products of oxidation of methyl ethyl sulfide under standard conditions and at a different temperature, methods used to calculate thermodynamic parameters including enthalpy of reactions, the entropy of reactions and Gibbs free energy of reactions. It also presents the listing of thermodynamic tables for literature values for the enthalpy of reactions, the entropy of reactions, and Gibbs free energy of reactions for all possible reaction pathways.

2. Introduction

Methyl ethyl sulfides are used in organic synthesis and scientific research. These compounds have a negative effect on the atmosphere and the environment by generating sulfuric oxides.

Understanding the oxidation of sulfur compounds is important for the reduction and prevention of sulfur oxides, as they are harmful to human health, building, and the environment.

Due to biological activity in the sea, a significant amount of ethyl methyl sulfide is emitted, and the concentration of ethyl methyl sulfide in the atmosphere is the result of anthropogenic and human activity. Methyl ethyl sulfides have a negative effect on the atmosphere as it produces sulfuric oxides in the environment. Recent studies show that the partial oxidation of methyl ethyl sulfide begins with the abstraction of a hydrogen atom from each of its carbon sites.

$$CH_3SCH_2CH_3 + HO \cdot \rightarrow CH_2 \cdot SCH_2CH_3$$

$$CH_3SCH_2CH_3 + HO \cdot \rightarrow CH_3SCH \cdot CH_3$$

$$CH_3SCH_2CH_3 + HO \cdot \rightarrow CH_3SCH_2CH_2 \cdot$$

The unstable radical products would subsequently be oxidized and form unstable alkyl sulfide proxy radicals.

$$CH_2 \cdot SCH_2CH_3 + O_2 \rightarrow \cdot OOCH_2SCH_2CH_3$$

$$CH_3SCH \cdot CH_3 + O_2 \rightarrow CH_3SCH(OO \cdot)CH_3$$

$$CH_3SCH_2CH_2 \cdot + O_2 \rightarrow CH_3SCH_2CH_2OO \cdot$$

The reaction pathway for oxidation of each radical, $CH_2 \cdot SCH_2CH_3$ 10, $CH_3SCH \cdot CH_3$ 11, and $CH_3SCH_2CH_2 \cdot$, has been studied.

The study of methyl ethyl sulfide oxidation products is of value to knowing the fate of its oxidation products in the atmosphere. In the past several studies were conducted to detect and identify the oxidation products of ethyl methyl sulfide at different reaction conditions.

In 2012, studies made by the Gabriela Oksdath-Mansilla group showed that hydroxyl radical initiated oxidation of ethyl methyl sulfide at 1 atm pressure and 298 K in a borosilicate glass photoreactor and quartz-glass photoreactor in synthetic air, yielded 51 ± 2 % SO2, 46 ± 4 % HCHO, 57 ± 3 % CH3CHO and 0.07% OCS. And the Photooxidation products by OH in the presence of oxygen at atmospheric pressure of diethyl sulfide are 50 ± 3 % SO2, 91 ± 3 % CH3CHO, and 0.14% OCS, and of dimethyl sulfide are 80 ± 10 % SO2 (Barnes et al. 1996), 90 ± 6 % HCHO, $0.7\% \pm 0.2$ %OCS, and 95 ± 4 % SO2 (Arsene at al., 1999).

In 2010, studies made by Zheng and Bozzelli showed that pyrolysis and oxidation of ethyl methyl sulfide at 630-740°C and 1 atm pressure in a turbulent flow reactor yielded ethylene, methane, and ethane at a significant level; other species were also formed and detected, carbon dioxide, carbon monoxide, formaldehyde, and sulfur dioxide.

Three thermodynamic quantities can be used to predict the spontaneity of a reaction, therefore, predict the favored products formed at a certain set of conditions, e.g., standard conditions 298 K and 1atm pressure; these thermodynamic quantities can also be used to predict favored products at different temperatures. These thermodynamic quantities are standard enthalpy change (ΔHrxn), Standard entropy change (ΔSrxn), and standard Gibbs free energy change (ΔGrxn) for a reaction. It's known that if the sign for both calculated standard enthalpy change and standard entropy change for a reaction are negative, regardless of the sign of calculated Gibbs free energy change for the reaction, the reaction would be spontaneous only at low temperatures. Therefore the reaction would favor the formation of products only at low temperatures. If the sign for both calculated standard enthalpy change and standard entropy change for a reaction are positive, regardless of the sign of calculated Gibbs free energy change for the reaction, the reaction would be spontaneous only at high temperatures. If the sign for the calculated standard enthalpy change for a reaction is negative, the sign for the calculated standard entropy change for a reaction is positive, and the sign for the calculated Gibbs free energy change for the reaction is negative, the reaction would be spontaneous at all temperatures. If the sign for the calculated standard enthalpy change for a reaction is positive, the sign for the calculated standard entropy change for a reaction is negative, and the sign for the calculated Gibbs free energy change for the reaction is positive, the reaction would be non-spontaneous at all temperatures. Therefore the reaction wouldn't favor the formation of products at all temperatures.

Enthalpy of reaction (H) is a thermodynamic quantity, and it's the heat involved in chemical or physical change at constant temperature and pressure., q=ΔH. The standard enthalpy change for a reaction (ΔH°_{rxn}) can be calculated using tabulated standard enthalpy of formation of species involved in the reaction at $25°C$ and 1 atm pressure,

$$\Delta H_{rxn}° = \Sigma n \, \Delta H_f° \, (\text{products}) - \Sigma m \, \Delta H_f° \, (\text{reactants}).$$

Entropy (S) is a measure of the entropy dispersal of a system, and its other is a thermodynamic quantity. In a spontaneous process, the entropy of the system and the surroundings would increase, and energy would be dispersed. Entropy change for a reaction ($\Delta S°_{rxn}$) can be calculated using standard entropy values of reactants and products,

$$\Delta S_{rxn}° = \Sigma n \, \Delta S° \, (\text{products}) - \Sigma m \, \Delta S° \, (\text{reactants}).$$

Standard free energy change ($\Delta G°$), it's a free energy change that occurs when products in their standard states are formed from reactants in their standard states. Standard states refer to 1 atm partial pressure for gases, 1M concentration for solutions, 1 atm pressure for solutions and liquids, and the temperature is $25°C$ or 298K. The standard free energy for a reaction can be calculated using the equation,

$$\Delta G°_r = \Delta H° - T \, \Delta S°.$$

Free radicals are ions, molecules, or atoms containing at least one unpaired electron in the valance shell. These unpaired valance electrons make radicals unstable and chemically reactive. Most radicals have short lifetimes. Radicals can be generated by ionizing radiation, electrical discharge, electrolysis, and heat. In many chemical reactions, free radicals are intermediates. Free radicals are important in many chemical processes, including atmospheric chemistry, combustion, polymerization, plasma chemistry, and biochemistry.

Thermochemical and kinetic studies of the reaction between the activated secondary ethyl radical of ethyl methyl sulfide ($CH_3SCH \cdot CH_3$) and molecular oxygen were analyzed by Song, Bozzelli, and Abdel-Wahab. Reaction enthalpies (ΔH_{rxn}) and energy barriers (E_a) and for the reactions of $CH_2SCH(OO \cdot)CH_3$ adduct under the M06-2X/6-311+G(2d,p), G3MP2B3 and CBS-QB3 calculation and possible products were studied. New possible reaction pathways with different calculated enthalpy of reactions were presented. It has been found that at temperatures between 600 ~ 800 K and 1 atm pressure, the reaction of the CH3SCH(OO•)CH3 adduct product would occur, at below 500K stabilization of peroxy adduct product would occur rather than the formation of products, and all of the subsequent reaction paths occur at the optimal temperature of 800 K.

Kinetics and thermodynamic studies can also provide insights into an adsorption process and its capacity. Adsorption processes are used in purification systems. Kinetics and thermodynamic studies of anionic dye Methyl Orange adsorbed by modified polyvinylidene fluoride electrospun fibers (polyvinylidene fluoride-PEDOT mats) in aqueous solutions showed that this compound could selectively remove methyl orange on a PH ranging from 3 to 10 with a maximum capacity of 143.8 mg/g in an acidic PH of 3. It's used in the recirculating filtration system, and the maximum absorption capacity has been found to be reached in a shorter time. The adsorption process was found to be endothermic and spontaneous by estimating entropy, Gibbs free energy, and enthalpy changes for the adsorption process. Also, the mechanism for the interaction between the membrane and methyl orange molecules are found to be electrostatic based on adsorption isotherm data and analysis of the kinetics.

3. Calculation Methods

Standard entropy and Standard Gibbs free energy change for possible pathways of oxidation of methyl ethyl sulfide radicals has been calculated using the method described in the introduction, and the calculated values are used to predict the spontaneity of reaction pathways and are also used to determine thermionically favored products at 25°C/ 298 K and 1 atm pressure.

4. Results and Analysis

The calculated enthalpy change for the reaction was comparable to the calculated enthalpy change for the reaction using the CBS-QB3 DFT method, with values differing in the range of 0.1- 3.3 Kcal/mol, indicating that the CBS-QBS DFT method is an accurate method in calculating enthalpy change for a reaction.

The standard entropy change for possible pathways of oxidation of methyl ethyl sulfide radicals has been calculated using thermochemical properties for species involved and previously calculated using the CBS-QB3 DFT method. The standard entropy change for reaction pathways 1-11 are 58, 41.5, 1.7, 20.9, 54.6, 20.7, 41.4, 27.7, 87.3, 55.6, and 53.3 cal/mol.K, table 1.

The standard Gibbs free energy change for possible pathways of oxidation of methyl ethyl sulfide radicals has been calculated using thermochemical properties for species involved and previously calculated using the CBS-QB3 DFT method. The standard Gibbs free energy change for reaction pathways 1-11 are 26.2, 15.3, -23.7, -55.9, -23.2, 1.4, 7.6, 37.9, -33.0, -8.0, and -69.2 kcal/mol, table 1.

For a reaction to be spontaneous and form an Energetically favorable product at all temperatures, the sign of both enthalpies change of the reaction and Gibbs free energy of the reaction must be negative, and the sign of entropy change of the reaction must be positive.

Reaction pathway numbers 3,4,5, 9, and 11 show a negative sign for both the calculated standard enthalpy and standard Gibbs free energy change for the reaction, but a positive sign for entropy change for reactions, therefore, these reaction pathways occur spontaneously at all temperatures, including standard conditions, and the reactions would favor the formation of products.

Reaction pathway numbers 1,2,6,7, and 8 show a positive sign for all calculated thermolytic properties; standard entropy, standard enthalpy, and standard Gibbs free energy change for the reaction pathways; therefore, these reaction pathways are spontaneous only at high temperatures. These reaction pathways are nonspontaneous at standard conditions. Reaction pathway numbers 10 show a positive sign for the calculated standard entropy and standard enthalpy change for the reaction, but a negative sign for the calculated standard Gibbs free energy change for the reaction. Therefore, this reaction pathway is also spontaneously only at high temperatures, and non-spontaneous at standard conditions.

The Energetically favorable oxidation products formed at standard conditions, 1 atm pressure, and 298K, are Acetaldehyde ($CH_3CH=O$), Thioformaldehyde ($CH_2=S$), Thio-acetaldehyde ($CH_3CH=S$), Formaldehyde ($CH_2=O$). And some short-lived, unstable, and reactive radicals, $CH_3S\bullet=O$, $CH_3CH_2S\bullet=O$, $CH_3S(=O)CH_2CO\bullet$, and OH are also formed under standard conditions.

Two of these products, Acetaldehyde ($CH_3CH=O$) and Formaldehyde ($CH_2=O$), were identified as major photooxidation products of ethyl methyl sulfide at atmospheric pressure; it was found that 46 ± 4 % of $CH_2=O$ and 57 ± 3 % of $CH_3CH=O$ was present in the reaction mixture in 2012. 7.

The Energetically favorable oxidation products formed at 1 atm pressure and at high temperatures are Ethylene (CH2=CH2), Thioformaldehyde (CH2=S), S=CHCOOH, and CH2=CHOOH And some short-lived stable radicals, HO2, CH3•, HOOCH2S •, and CH3S• are also formed at high temperatures.

Two of these products, Ethylene (CH2=CH2) and water (H2O), were identified as products of pyrolysis and oxidation of ethyl methyl sulfide at 630-740° C and at 1 atm pressure in 2010.

Table1. Calculated Entropy and Gibbs free energy change for all possible pathways of oxidation of methyl ethyl sulfide radicals

Number	Reaction Pathway	CBS-QB3 $\Delta Hrxn^a$	Calculated $\Delta Hrxn^a$	Calculated $\Delta Srxn^b$	Calculated $\Delta Grxn^a$
Reaction 1	$CH_3SCH_2CH_2 \bullet + O_2 \rightarrow CH_3SCH_2CH_2OO \bullet$ CH2•SCH2COOH $\rightarrow$ $\rightarrow$ CH2=CH2 + HO2 + CH2=S	28.2	31.5	58	26.2
2	CH3SCH•COOH $\rightarrow$ CH3• + S=CHCOOH	27.7	27.2	41.5	15.3
3	CH3SCH2COO• $\rightarrow$ CH3S(=O)CH2CO•	-23.2	-23.1	1.7	-23.7
Reaction	$CH_2\bullet SCH_2CH_3 + O_2 \rightarrow \bullet OOCH_2SCH_2CH_3$				

#	Reaction				
4	•OOCH2SCH2CH3 → CH 2 = O + CH 3 CH 2 S • = O	-49.7	-49.7	20.9	-55.9
5	HOOCH2SCH•CH3 → CH2CH = S + CH 2 = O + OH	-6.9	-6.9	54.6	-23.2
6	HOOCH2SCH2CH2• → HOOCH2S • + CH2 = CH2	9.0	9.0	20.7	1.4
7	HOOCH2S • → HO 2 + CH 2 = S	19.9	20.9	41.4	7.6
8	•OOCSCC → •CH 2 •OO • + CH 3 CH 2 S •	46.2	47.8	27.7	37.9
Reaction	$CH_3SCH•CH_3 + O_2 \rightarrow CH_3SCH(OO •)CH_3$				
9	CH2•SCH(OOH)CH3 → CH2=S + CH3CH=O + OH	-7.0	-7.4	87.3	-33.0
10	CH3SCH(OOH)CH2• → CH3S•+ CH2=CHOOH	8.5	8.5	55.6	-8.0
11	CH3SCH(OO•)CH3 → CH3CH=O + CH3S•=O	-53.3	-53.3	53.3	-69.2

[a] **kcal/mol**

[b] **cal/mol.K**

5. Summary and Conclusion

The listing of thermodynamic tables for literature values for the enthalpy of reactions, the entropy of reactions, and Gibbs free energy of reactions for all possible reaction pathways are included in table 1. Three thermodynamic quantities used to predict the spontaneous reaction pathways and the reaction pathways that energetically favor products at standard conditions, Gibbs free change for reaction, the standard enthalpy change for reaction, and standard entropy change for the reaction are presented in the introduction.

The energetically favorable stable products formed from oxidation of ethyl methyl sulfide at standard conditions, 298 K and 1 atm pressure are acetaldehyde ($CH_3CH=O$), thioformaldehyde ($CH_2=S$), thio-acetaldehyde ($CH_3CH=S$), formaldehyde ($CH_2=O$).

The favored reactive and unstable radicals formed under standard conditions are $CH_3S•=O$, $CH_3CH_2S•= O$, $CH_3S(=O)CCO•$, and OH. The Energetically favorable stable products formed from the oxidation of ethyl methyl sulfide at high temperatures and 1 atm pressure are Ethylene (CH2=CH2), Thioformaldehyde (CH2=S), S=CHCOOH, and CH2=CHOOH, the favored reactive and unstable radicals formed at high temperatures and 1 atm pressure are HO2, CH3•, HOOCH2S • and CH3S•.

Chapter 7
Enthalpy, Entropy, and Heat Capacity of some Fluorinated Ethanol's and its Radicals at Different Temperatures

1. Overview

The thermochemical properties of fluorinated alcohols are needed to understand their stability and reactions in the environment and in the thermal process.

This chapter presents the use of gaussian M-062x/6-31 + g (d, p) Ab initio and density functional theory method to calculate some kinetic and thermodynamic parameters for simple fluorinated organic alcohols: enthalpies of formation values in Kcal/mol, entropy and heat capacity values in cal/mol.K, the internal rotor contribution of each molecule are incorporated to the calculated values of entropy and heat capacity to obtain more accurate calculated values. This chapter also includes the optimized structures (fig1-19) and thermochemical properties of the species determined by the Gaussian M-062x/6-31+g (d,p) method, contributions of entropy, $S°298$, and heat capacities, $Cp(T)$ due to vibration, translation, and external rotation of the molecules using the vibration frequencies and structures obtained from the M-062x/6-31+g (d,p) density functional method. Calculated enthalpies of formation for 19 fluorinated ethanol and some radicals with a popular ab initio and density functional theory methods are also presented: via several series of isodesmic reactions. Entropies ($S298°$ in cal mol^{-1} K^{-1}) were estimated using the M-062x/6-31+g (d,p) computed frequencies and geometries. Rotational barriers hindered internal rotational contributions for S298 °- 1500°, and $Cp(T)$ using the rigid rotor harmonic oscillator approximation, with direct integration over energy levels of the intramolecular rotation potential energy curves, are also presented. Tabulated literature values for entropy, and heat capacity at different temperatures are also included.

2. Introduction

Fluorinated hydrocarbons are used as refrigerants, in polymers, in heat exchange fluids, and as solvents. They are present in the atmosphere, lithosphere, and hydrosphere. Because of their less adverse effects on the stratospheric ozone layer, they are used in place of greenhouse gases. Fluorinated hydrocarbons exist as compounds ranging from pure to oxidized intermediates resulting from oxidation in the environment. In order to study their reactivity in biological systems, lifetimes, and the environment, it's critical to understand the chemical and thermodynamic properties of fluorocarbons and their breakdown intermediates.

The thermochemistry of fluorinated alcohols with one carbon atom was studied in the past and is in the literature. In 2016 Hang Wang studied the thermodynamic properties of fluorinated methanol using CBS-QB3, M06, M06-2X, WB97X, W1U, B3LYP, CBS-APNO, and G4 Calculations. A small standard deviation suggests good error cancellation of work reactions and accuracy. M06-2x/6-31+g (d,p) calculation had small values for standard deviations; it is an accurate method to calculate the Enthalpy of fluorinated alcohols; it shows the second smallest standard deviation after CBS-QB3 method of calculation

Halogenated compounds are highly stable, have low reactivity, and are valued chemicals in industry. Due to their widespread use and their persistence in the environment, they are of concern to the environment. In order to understand the oxidation and reduction reactions involving such molecules, their thermochemical properties must be studied.

The thermochemistry of fluorinated alcohols with one carbon atom was studied in the past and is in the literature. In 2016 Hang Wang studied the thermodynamic properties of fluorinated methanol using CBS-QB3, M06, M06-2X, WB97X, W1U, B3LYP, CBS-APNO, and G4 Calculations. A small standard deviation suggests good error cancellation of work reactions and accuracy. M06-2x/6-31+g (d,p) calculation had small values for standard deviations; it is an accurate method to calculate the Enthalpy of fluorinated alcohols, and it shows the second smallest standard deviation after the CBS-QB3 method of calculation.

The onset temperatures for energetic materials in the calorimetric measurements have been roughly predicted using molecular orbital calculations of bond dissociation energies. The stability and reactivity of chemical compounds can be explained using bond dissociation energies values. Standard enthalpies of formation estimated using semi-empirical MO calculations, the MOPAC-PM7 package has been used previously to derive bond dissociation energy values for chemical compounds.

Bond dissociation energies (BDEs) values for some inorganic compounds, lanthanide selenides, and sulfides were measured in 2021 using Resonant two-photon ionization spectroscopy. The pre-dissociation thresholds were found to be the BDE values for these molecules. The 0 K gaseous heat of formation, $\Delta_f H\circ K$ for each molecule, was also reported using this method.

3. Experimental and Data

3.1 Computational Method

Composite calculations and a series of Isodesmic reactions used to calculate the enthalpy of the formation of fluorinated ethanols are presented in summary, table 2 and, in detail, table 4. All calculations are performed using the Gaussian 16 program. The M-062x/6-31+g (d,p) level of calculation has been applied to fluoro hydrocarbons with small reported standard deviation values in the past.

The DFT method M06-2x is used to initially analyze optimized structures, frequencies, thermo energies, and internal rotors of the molecules studied. It's a Global-hybrid meta-GGA density functional approximation, GGA, generalized gradient approximation, where the density functional depends on the up and down spin densities and their reduced gradient, meta GGA, where the functional also depends on the up and down spin kinetic energy densities, hybrid GGA, a combination of GGA with Hartree-Fock exchange, hybrid meta GGA, a combination of meta GGA with Hartree-Fock exchange. All reported values are for the standard state of 298 K and 1 atm.

3.2 Isodesmic and Isogyric Reaction

The Gaussian M-062x/6-31+g (d,p) calculated enthalpy of formation of mono and di fluorinated ethanol and its radicals are presented in table 4. Work reactions along with reference species have been utilized to calculate the enthalpy of formation of fluorinated ethanols using this method.

The number of each type of bond must be conserved in the isodesmic reactions in order to cancel any systematic error in the molecular orbital calculations using this method. Calculations of enthalpies of formation are allowed to accuracies close to experimental values by the careful choice of the isodesmic reactions1. Taking 1-fluoroethanol as an example, two isodesmic reactions (Table 2) are selected to determine the $\Delta Hf298\,°$ of the target molecule, 1-fluoroethanol. Since the $\Delta Hf298\,°$ values of all species but 1-fluoroethanol in $1-2$ (Table 2) are known, the $\Delta Hf298\,°$ of the target species 1-fluoroethanol, is obtained from this data and the calculated $\Delta Hrxn, 298\,°$. $\Delta Hf298\,°$ calculated using two different reference molecules are within $\pm\,.2$ Kcal mol^{-1}.

3.3 Reference Species

Standard enthalpy of formation for the reference species used in the isodesmic work reactions, along with their uncertainties, are presented in table 1. Table 2 provides one example of the method of Isodesmic Work Reactions used for the calculation of the Standard Enthalpy of Formation $\Delta_f H°_{(298)}$ for the fluoroethanols.

Table 1. Reference Species in the Isodesmic Reactions Standard Enthalpy of Formation Values (kcal mol−1)

species	$\Delta_f H^\circ_{298}$	species	$\Delta_f H^\circ_{298}$
CH_3F	−56.54 ± 0.07[a]	CH_3OOH	−30.96 ± 0.67[b]
	−56.62 ± 0.48[h]	CH_3CH_2OOH	−38.94 ± 0.81[b]
CH_3CH_2F	−65.42 ± 1.11[a]	$CH_3CH_2CH_2OOH$	−44.03 ± 0.67[b]
$CH_3CH_2CH_2F$	−70.24 ± 1.30[a]	$CH_3OO^\bullet$	2.37 ± 1.24[b]
CH_2F_2	−108.07 ± 1.46[a]	$CH_3CH_2OO^\bullet$	−6.19 ± 0.92[b]
	−107.67 ± 0.48[h]	$CH_3CH_2CH_2OO^\bullet$	−11.35 ± 1.24[b]
CH_3CHF_2	−120.87 ± 1.62[a]	CH_4	−17.81 ± 0.01[c]
$CH_3CH_2CHF_2$	−125.82 ± 1.65[a]	CH_3CH_3	−20.05 ± 0.04[c]
CHF_3	−166.71 ± 1.97[a]	$CH_3CH_2CH_3$	−25.01 ± 0.06[i]
	−166.09 ± 0.48[h]	$CH_3CH_2CH_2CH_3$	−30.07 ± 0.08[i]
CH_3CF_3	−180.51 ± 2.05[a]	$CH_3O^\bullet$	5.15 ± 0.08[c]
$CH_3CH_2CF_3$	−185.48 ± 2.15[a]	$CH_3CH_2O^\bullet$	−3.01[d]
$CH_3^\bullet$	34.98 ± 0.02[c]	OH	8.96 ± 0.01[c]
$CH_3CH_2^\bullet$	28.65 ± 0.07[c]	CH_3OH	−47.97 ± 0.04[c]
$CH_3CH_2CH_2^\bullet$	24.21 ± 0.24[g,j]	CH_3CH_2OH	−56.07 ± 0.05[i]
	24.18[i]	$HOO^\bullet$	2.94[e,j]
H	52.10[c]	HOOH	−32.39 ± 0.04[f,j]
O	59.57[c]		−32.37[i]

Table 2. Example for Enthalpy of Formation Calculations for 1-fluoroethanol using Isodesmic Reactions; using different reference molecules, Units in kcal mol⁻¹

Isodesmic Reactions	ΔH°_{Rxn} (298) Hartrees	$\Delta H_{Rxn\ (298)}$ kcal/mol⁻¹	$\Delta_f H^\circ_{(298)}$ CH3−xCH2FxOH, CH3CH2-xFxOH kcal mol⁻¹	Error kcal mole⁻¹	Equ. #
CH2FCH2OH + CH₄ = CH3CH2OH + CH3F -254.173115[a] -40.447961 -154.926666 -139.683801 Reference Values -17.81 -56.21 -56.54 Kcal mol-¹	0.010609	6.657243	-101.6	±0.2	1
CH2FCH2OH + CH₃CH₃ = CH3CH2OH + CH3CH2F -254.173115 -79.717768 -154.926666 -178.963776 -20.05 -56.21 -65.42	0.000441	0.276731	-101.9	±1.3	2
Reported $\Delta_f H^\circ$ (298) kcal mol⁻¹			**-101.7 ± 0.7**		
Standard Deviation over rxns			**±0.1**		

a Hartrees, kcal mole⁻¹   *SD Standard Deviation kcal mol⁻¹ **Errors reported avg of sum of uncertainties in rxn's reference species**

4. Results and Discussion

4.1 Entropy and Heat Capacity

Internal rotor contributions to calculated entropy and heat capacity at 298-1500K were determined using the molecular mass of each molecule, the number of optical isomers, symmetry of the molecule, electron degeneracy, moment of inertia, and vibrational frequencies values (table 5). The vibrational frequencies for the calculation of heat capacity and entropy at the M-062x/6-31+g (d,p) level of calculation were scaled by a factor of 0.97 [7]. The moment of inertia values is shown in the <u>Supporting Information Table provided</u>. To calculate the contributions of the external rotor, vibration, and transition to the calculated entropy and heat capacity, the "SMCPS" 8 programs are used. It employs the rigid-rotor harmonic oscillator approximation using the moment of inertia from optimized structure and frequencies. The "Rotator" 9 programs by Lay et al.[10] are used to calculate internal rotor contributions from the corresponding internal rotor torsion frequencies. In this paper, a torsional potential curve presenting a ten-parameter Fourier series function is used to calculate the contribution of the internal rotor. Parameters and detailed functions are shown in the <u>Supporting Information Table provided</u>. Rotor 11 program is used to calculate thermodynamic functions from hindered rotations with arbitrary potentials.

Calculation of the Hamiltonian matrix of the internal rotor and subsequent calculation of energy levels by direct diagonalization of the matrix are employed by this technique. Rotational barriers versus dihedral angles are presented as a potential curve. In this paper, the calculated torsional potential at discrete torsional angles

$$V(\emptyset) = a_0 + \sum_{i=1}^{10} a_0 \cos(i\emptyset) + \sum_{j=1}^{10} b_j \cos(j\emptyset)$$

Equation 1

The coefficients ai, and bj are calculated to present the maxima and minima of the torsional potentials with a possibility to shift from the extreme angular positions.

Calculations of heat capacity and standard entropy based on the benchmark database and the computational chemistry comparison for the M-062x/6-31+g (d,p) calculation method, the vibrational frequencies were scaled by a factor of 0.987[12]. Potential Energy profiles for mono and di-fluorinated ethanol and their related radicals are listed in the <u>Supporting Information Table provided</u>. The solid lines are the fit of Fourier series expansion, rotator contribution for barriers below 7 kcal mol⁻¹ were added to the SMCPS calculated entropy, and heat capacity. Energies are in kcal mo⁻¹. Table 6.

Table 3. Monofluoro and Difluoro- Ethanol's Ideal Gas phase Entropy and Heat Capacity obtained using M-062x/6-31+g (d,p) level of theory (Cal mol⁻¹ K⁻¹)

SPECIES		S (298)	Cp (298)	Cp (400	Cp (500)	Cp (600)	Cp (800)	Cp (1000	Cp (1500)
CH2FCH2OH	TVR	64.74	13.23	17.12	20.72	23.75	28.18	31.11	35.06
	Internal	3.84	3.24	3.48	3.43	3.22	2.68	2.23	1.62
	Rotor	2.02	3.15	3.33	3.04	2.66	2.05	1.69	1.3
	Total	70.6	19.62	23.94	27.2	29.62	32.91	35.03	37.98
CH3CHFOH	TVR	65.17	14.06	17.9	21.37	24.26	28.51	31.32	35.15
	Internal	4.48	2.13	2.09	1.95	1.8	1.56	1.4	1.19
	Rotor	2.67	2.17	2.21	2.18	2.09	1.89	1.71	1.4
	Total	72.32	18.35	22.21	25.5	28.16	31.96	34.42	37.75
C•HFCH2OH	TVR	66.3	13.54	16.96	19.97	22.48	26.29	29.04	33.33
	Internal	6.19	2.31	1.91	1.69	1.54	1.36	1.25	1.13
	Rotor	3.09	3.16	2.52	2.04	1.74	1.41	1.25	1.11
	Total	75.58	19.01	21.38	23.7	25.76	29.06	31.55	35.56
CH2FCH•OH	TVR	66.47	13.76	17.09	20.04	22.52	26.29	29.02	33.31
	Internal	5.72	2.34	2.22	2.09	1.95	1.73	1.56	1.32
	Rotor	6	1.96	1.95	1.95	1.91	1.79	1.65	1.4
	Total	78.19	18.06	21.27	24.07	26.38	29.81	32.24	36.03
CH2FCH2O•	TVR	75.3	20.44	23.66	26.27	28.39	31.55	33.79	37.16
	Internal Rotor	6.08	2.2	1.96	1.7982	1.68	1.53	1.43	1.27
	Total	81.38	22.64	25.62	28.07	30.07	33.08	35.21	38.42

CH2•CHFOH	TVR	66.89	14.8	18.21	21.07	23.4	26.9	29.44	33.49
	Internal	4.94	1.42	1.25	1.17	1.12	1.06	1.04	1.01
	Rotor	2.84	1.95	2.12	2.12	2.04	1.83	1.64	1.35
	Total	74.67	18.17	21.58	24.35	26.55	29.79	32.11	35.86
CH3CF•OH	TVR	66.56	13.94	17.25	20.17	22.62	26.38	29.11	33.38
	Internal	4.91	1.97	1.75	1.57	1.44	1.27	1.18	1.08
	Rotor	1.71	2.6	3.17	3.22	3.01	2.43	1.99	1.45
	Total	73.18	18.5	22.17	24.96	27.07	30.08	32.28	35.92
CH3CHFÒ•	TVR	66.73	14.44	18.16	21.39	24.09	28.18	31.12	35.58
	Internal Rotor	4.55	2.12	2.05	1.9	1.75	1.51	1.36	1.18
	Total	71.27	16.56	20.21	23.29	25.84	29.69	32.48	36.75
CH2FCHFOH	TVR	67.46	15.41	19.51	23.12	26.07	30.23	32.83	36.15
	Internal	0	0	0	0	0	0	0	0
	Rotor	6.12	3.11	2.64	2.24	1.94	1.57	1.37	1.17
	Total	73.58	18.52	22.15	25.35	28	31.79	34.2	37.31
CF2HCH2OH	TVR	62.94	12	15.54	18.95	21.9	26.43	29.57	34.06
	Internal	4.98	3.36	3.2	2.97	2.72	2.26	1.93	1.48
	Rotor	2.23	3.21	3.17	2.81	2.44	1.91	1.6	1.27
	Total	70.15	18.57	21.91	24.72	27.06	30.6	33.1	36.81
CH3CF2OH	TVR	68.41	16.53	20.74	24.23	27.01	30.87	33.29	36.37
	Internal	4.53	2.13	2.08	1.93	1.77	1.53	1.38	1.18
	Rotor	1.58	2.45	3.09	3.28	3.14	2.6	2.14	1.53
	Total	74.52	21.11	25.91	29.44	31.93	35.01	36.8	39.08
CH2FC•FOH	TVR	69.58	15.15	18.76	21.84	24.36	28.05	30.59	34.38
	Internal	6.18	3.28	2.51	2.07	1.8	1.5	1.34	1.16
	Rotor	3.15	1.91	1.96	1.91	1.83	1.65	1.5	1.27
	Total	78.9	20.34	23.22	25.82	27.99	31.19	33.43	36.81
C•HFCFHOH	TVR	69.6	15.48	19.22	22.3	24.76	28.32	30.75	34.41
		6.92	2.03	1.69	1.49	1.37	1.22	1.15	1.07

	Internal Rotor	1.72	2.79	3.1	3.03	2.83	2.37	2.02	1.52
	Total	78.25	20.3	24.01	26.83	28.96	31.91	33.92	37
CH2FCHFO•	TVR	70.53	15.99	19.88	23.22	25.94	29.93	32.65	36.59
	Internal Rotor	5.48	2.92	2.79	2.58	2.37	2.01	1.76	1.4
	Total	76.01	18.91	22.67	25.8	28.31	31.94	34.41	37.99
CHF2C•HOH	TVR	68.27	15.46	19.18	22.25	24.71	28.28	30.72	34.4
	Internal	6.07	2.65	2.3	2.05	1.86	1.61	1.45	1.24
	Rotor	1.51	2.5	3.06	3.18	3.06	2.6	2.2	1.61
	Total	75.85	20.61	24.54	27.48	29.63	32.49	34.37	37.25
C•F2CH2OH	TVR	69.57	15.16	18.73	21.81	24.32	28.03	30.58	34.37
	Internal	6.18	2.68	2.38	2.11	1.89	1.5848	1.41	1.19
	Rotor	3.46	2.45	2.04	1.75	1.55	1.32	1.21	1.09
	Total	79.21	20.29	23.15	25.66	27.76	30.94	33.19	36.65
CHF2CH2O•	TVR	70.02	15.31	18.91	22.14	24.87	29.03	31.94	36.19
	Internal Rotor	5.78	2.51	2.41	2.26	2.1	1.83	1.63	1.34
	Total	75.79	17.82	21.32	24.4	26.97	30.86	33.57	37.53
CH2•CF2OH	TVR	70.2	17.3	21.07	23.93	26.13	29.24	31.38	34.69
	Internal	4.97	1.37	1.23	1.16	1.11	1.06	1.04	1.01
	Rotor	1.78	2.81	3.33	3.25	2.9407	2.3	1.87	1.38
	Total	76.95	21.48	25.63	28.35	27.25	32.61	34.29	37.09
CH3CF2O•	TVR	70.14	16.85	20.78	23.99	26.55	30.3	32.87	36.66
	Internal	4.84	2.08	1.89	1.7	1.54	1.35	1.23	1.11
	Rotor	5.78	2.51	2.41	2.26	2.1	1.83	1.63	1.34
	Total	80.75	21.44	25.08	27.94	30.2	33.47	35.73	39.11

4.2 Standard Enthalpy

Enthalpies of formation from isodesmic work reactions from the M-062x/6-31+g (d,p) method of calculation are presented in table 4. The standard enthalpy of formation for the reference species, along with their uncertainties, which are used in the isodesmic work reactions, table 4, are listed in table 1 in kcal mol^{-1}. The standard deviation was calculated for all Enthalpies of formation values for all 19 fluorinated ethanol and are included in table 4. Details of the method of standard deviation and example calculation(s) are shown in the *Supporting Information table provided.

Table 4. Standard Enthalpy of Formation using isodesmic reactions: Monofluoro and Difluoro- Ethanols using the M06-2x/6-31+g (d,p) Level of Theory.

Isodesmic Reactions Target Specie	ΔH° Rxn (298) Hartrees	ΔH° Rxn (298) Kcal/mol e^1	$\Delta_f H^\circ$ (298) kcal mol^{-1}	Error kcal mol^{-1}
CH2FCH2OH + CH$_4$ = CH3CH2OH + CH3F -254.173115 -40.447961 -154.926666 -139.683801 -17.81 -56.21 -56.54	0.010609	6.657243	**-101.6**	±0.2
CH2FCH2OH + CH$_3$CH$_3$ = CH3CH2OH + CH3CH2F -254.173115 -79.717768 -154.926666 -178.963776 -20.05 -56.21 -65.42	0.000441	0.276731	**-101.9**	±1.3
Reported Δ_fH° (298) kcal mol^{-1}			**-101.7 ± 0.7**	
Standard Deviation over rxns			**±0.1**	
CH3CHFOH + CH3CH2CH3 = CH3CH2OH + CH3CH2CH2F -254.19203 -118.990915 -154.926666 -218.237381 -25.02 -56.21 -70.24	0.018898	11.85867	**-113.3**	±1.5
CH3CHFOH + CH3CH3 = CH3CH2OH + CH3CH2F -254.19203 -79.717768 -154.926666 -178.963776 -20.05 -56.21 -65.42	0.019356	12.14606	**-113.7**	±1.3
Reported Δ_fH° (298) kcal mol^{-1}			**-113.5 ±1.4**	
Standard Deviation over rxns			**±0.2**	
C•HFCH2OH + CH3CH2CH3 = CH3CH2O• + CH3CH2CH2F -253.518298 -118.990915 -154.268107 -218.237381 -25.02 -3.01 -70.24	0.003725	2.337471	**-50.6**	±1.4
C•HFCH2OH + CH4 = CH3CH2O• + CH3F -253.518298 -40.447961 -154.268107 -139.683801 -17.81 -3.01 -56.54	0.014351	9.005382	**-50.8**	±0.1

Reported $\Delta_f H^\circ$ (298) kcal mol^{-1}				
			-50.7 ± 0.8	
Standard Deviation over rxns			±0.1	
CH2FCH•OH + CH3CH3 = CH3CH2O• + CH3CH2F -253.518298 -79.717768 -154.268107 -178.963776 -20.05 -3.01 -65.42	0.012436	7.803702	-56.2	±1.2
CH2FCH•OH + CH4 = CH3CH2O• + CH3F -253.518298 -40.447961 -154.268107 -139.683801 -17.81 -3.01 -56.54	0.022604	14.18421	-55.9	±0.1
Reported $\Delta_f H^\circ$ (298) kcal mol^{-1}			- 56.1±0.6	
Standard Deviation over rxns			±0.1	
CH2FCH2O• + CH3CH2CH3 = CH3CHO• + CH3CH2CH2F -253.509195 -118.990915 -154.268107 -218.237381 -25.02 -3.01 -70.24	-0.00538	-3.37474	-44.9	±1.4
CH2FCH2O• + CH4 = CH3O• + CH3CH2F -253.509195 -40.447961 -114.989112 -178.963776 -17.81 5.15 -65.42	0.004268	2.678208	-45.1	±1.2
Reported $\Delta_f H^\circ$ (298) kcal mol^{-1}			- 45.0±1.3	
Standard Deviation over rxns			±0.1	
CH2•CHFOH + CH4 =CH3O• + CH3CH2F -253.532134 -40.447961 -114.989112 -178.963776 -17.81 5.15 -65.42	0.027207	17.07264	-59.5	±1.2
CH2•CHFOH + CH3CH3 = CH3CH2O• + CH3CH2F -253.532134 -79.717768 -154.268107 -178.963776 -20.05 -3.01 -65.42	0.018019	11.30708	-59.7	±1.2
Reported $\Delta_f H^\circ$ (298) kcal mol^{-1}			- 59.6±1.2	
Standard Deviation over rxns			±0.1	
CH3CF•OH + CH3CH3 = CH3CH2O• + CH3CH2F -253.542527 -79.717768 -154.268107 -178.963776 -20.05 -5.01 -65.42	0.028412	17.82879	-68.2	±1.2
CH3CF•OH + CH3CH2CH3 = CH3CH2O• + CH3CH2CH2F -253.542527 -118.990915 -154.268107 -218.237381 -25.02 -5.01 -70.24	0.027954	17.54139	-67.8	±1.4
Reported $\Delta_f H^\circ$ (298) kcal mol^{-1}			-68.0± 1.3	
Standard Deviation over rxns			±0.2	
CH3CHFO• + CH4 = CH3O• + CH3CH2F -253.530783 -40.447961 -114.989112 -178.963776 -17.81 **5.15** -65.42	0.025856	16.22487	-58.7	±1.2
CH3CHFO• + CH3CH3 = CH3CH2O• + CH3CH2F -253.530783 -79.717768 -154.268107 -178.963776 -20.05 -3.01 -65.42	0.016668	10.45932	-58.8	±1.2
Reported $\Delta_f H^\circ$ (298) kcal mol^{-1}			- 58.8±1.2	
Standard Deviation over rxns			±0.2	
CH2FCHFOH + CH4 = CH3CH2OH + CH2F2 -353.429924 -40.447961 -154.926666 -238.938973 -17.81 -56.21 -108.07	0.012246	7.684475	-154.2	±1.6

CH2FCHFOH +CH3CH2CH3 = CH3CH2OH + CH3CH2CHF2 -353.429924 -118.990915 -154.926666 -317.498882 -25.02 -56.21 -125.82	-0.00471	-2.95494	-154.1	±1.9
Reported $\Delta_f H^o$ (298) kcal mol^{-1}			-154.2±1.7	
Standard Deviation over rxns			±0.6	
CF2HCH2OH + CH4 = CH3CH2OH + CH2F2 -353.431533 -40.447961 -154.926666 -238.938973 -17.81 -56.21 -108.07	0.013855	8.694137	-155.2	±1.6
CF2HCH2OH + CH3CH3 = CH3CH2OH + CH3CHF2 -353.431533 -79.717768 -154.926666 -278.225303 -20.05 -56.21 -120.87	-0.00267	-1.67419	-155.4	±1.8
Reported $\Delta_f H^o$ (298) kcal mol^{-1}			-155.3±1.7	
Standard Deviation over rxns			±0.1	
CH3CF2OH + CH4 = CH3CH2OH + CH2F2 -353.462472 -40.447961 --154.926666 -238.938973 -17.81 -56.21 -108.07	0.044794	28.10864	-174.6	±1.6
CH3CF2OH + CH3CH2CH3 = CH3CH2OH + CH3CH2CHF2 -353.462472 -118.990915 -154.926666 -317.498882 -25.02 -56.21 -125.82	0.027839	17.46922	-174.5	±1.5
Reported $\Delta_f H^o$ (298) kcal mol^{-1}			-174.5±1.5	
Standard Deviation over rxns			±0.1	
CH2FC•FOH + CH4 = CH3CH2O• + CH2F2 -352.776406 -40.447961 -154.268107 -238.938973 -17.81 -3.01 -108.07	0.017287	10.84775	-104.1	±1.5
CH2FC•FOH + CH3CH2CH3 = CH3CH2O• + CH3CH2CHF2 -352.776406 -118.990915 -154.268107 -317.498882 -25.02 -3.01 -125.82	0.000332	0.208333	-104.0	±1.4
Reported $\Delta_f H^o$ (298) kcal mol^{-1}			-104.1 ± 1.5	
Standard Deviation over rxns			±0.1	
C•HFCFHOH + CH4 = CH3O• + CH3CHF2 -352.778474 -40.447961 -114.989112 -278.225303 -17.81 5.15 -120.87	0.01202	7.542658	-105.5	±1.7
C•HFCFHOH + CH3CH3 = CH3O• + CH3CH2CHF2 -352.778474 - 79.717768 -114.989112 -317.498882 -20.05 5.15 -125.82	0.008248	5.175694	-105.8	±1.8
Reported $\Delta_f H^o$ (298) kcal mol^{-1}			-105.6±1.7	
Standard Deviation over rxns			±0.2	
CHF2C•HOH + CH4 = CH3O• + CH3CHF2 -352.773245 -40.447961 -114.989112 -278.225303 -17.81 5.15 -120.87	0.006791	4.261414	-102.5	±1.7
CHF2C•HOH + CH3CH3 = CH3O• + CH3CH2CHF2 -352.773245 -79.717768 -114.989112 -317.498882 -20.05 5.15 -125.82		1.89445	-102.2	±1.8

	0.003019			
Reported $\Delta_f H^o$ (298) kcal mol^{-1}			-102.3±1.7	
Standard Deviation over rxns			±0.2	
CH2FCHFO• + CH4 = CH3CH2O• + CH2F2 -352.768218 - 40.447961 -154.268107 -238.938973 -17.81 -3.01 -108.07	0.009099	5.709704	-99.0	±1.5
CH2FCHFO• + CH3CH3 = CH3CH2O• + CH3CHF2 -352.768218 -79.717768 -154.268107 -278.225303 -20.05 -3.01 -120.87	-0.00742	-4.65863	-99.2	±1.7
Reported $\Delta_f H^o$ (298) kcal mol^{-1}			-99.1±1.6	
Standard Deviation over rxns			±0.1	
C•F2CH2OH + CH4 = CH3CH2O• + CH2F2 -352.773245 -40.447961 -154.268107 -238.938973 -17.81 -3.01 -108.07	0.014126	8.864192	-102.1	±1.5
C•F2CH2OH + CH3CH3 = CH3CH2O• + CH3CHF2 -352.773245 -79.717768 -154.268107 -278.225303 -20.05 -3.01 -120.87	-0.0024	-1.50414	-102.3	±1.7
Reported $\Delta_f H^o$ (298) kcal mol^{-1}			-102.2±1.6	
Standard Deviation over rxns			±0.2	
CHF2CH2O• + CH4 = CH3CH2O• + CH2F2 -352.767873 -40.447961 -154.268107 -238.938973 -17.81 -3.01 -108.07	0.008754	5.493214	-98.8	±1.5
CHF2CH2O• + CH3CH3 = CH3CH2O• + CH3CHF2 -352.767873 -79.717768 -154.268107 -278.225303 -20.05 -3.01 -120.87	-0.00777	-4.87512	-99.0	1.7
Reported $\Delta_f H^o$ (298) kcal mol^{-1}			-98.9±1.6	
Standard Deviation over rxns			±0.2	
CH2•CF2OH + CH4 = CH3O• + CH3CHF2 -352.800434 -40.447961 -114.989112 -278.225303 -17.81 5.15 -120.87	0.03398	21.32276	-119.2	±1.7
CH2•CF2OH + CH3CH3 = CH3O• + CH3CH2CHF2 -352.800434 -79.717768 -114.989112 -317.498882 -20.05 5.15 -125.82	0.030208	18.95579	-119.6	±1.8
Reported $\Delta_f H^o$ (298) kcal mol^{-1}			-119.4±1.7	
Standard Deviation over rxns			±0.2	
CH3CF2O• + CH4 = CH3CH2O• + CH2F2 -352.786755 -40.447961 -154.268107 -238.938973 -17.81 5.15 -108.07	0.027636	17.34184	-110.6	±1.5
CH3CF2O• + CH3CH2CH3 = CH3CH2O• + CH3CH2CHF2 -352.786755 -118.990915 -154.268107 -317.498882 -25.02 5.15 -125.82	0.010681	6.702424	-110.5	±1.8

Reported $\Delta_f H°$ (298) kcal mol[-1]			-110.6±1.6	
Standard Deviation over rxns			±0.1	

Errors reported as the sum of avg uncertainty in rxn's reference specie

Hartrees, kcal mole[-1] *SD Standard Deviation kcal mol[-1] **Errors reported avg of the sum of uncertainties in rxn's reference species**

4.3 Bond Energies

Table 5. Bond Dissociation Energy (BDEs) of Monofluoro and Difluoro-Ethanol

Reactions	Bond Dissociation Energy (Kcal mol[-1])	Error Kcal mol[-1]
	BDE (this study)	
<u>H-CHFCH2OH</u> H-CHFCH2OH = H• + •CHFCH2OH -101.7 ±0.1 52.1 -50.7±0.09	103.1± 0.1	±0.2
<u>CH2FC-HHOH</u> CH2FC-HHOH = H• + CH2FC•HOH -101.7 ±0.1 52.1 -56.1±0.1	97.7±0.1	±0.2
<u>CH2FCH2O-H</u> CH2FCH2O-H = H• + CH2FCH2O• -101.7 ±0.1 52.1 -45.0±0.1	108.8±0.1	±0.2
<u>H-CH2CHFOH</u> H-CH2CHFOH = H• + •CH2CHFOH -113.5 ±0.2 52.1 -59.6±0.1	106.0±0.15	±0.3
<u>CH3C-HFOH</u> H-CH2CHFOH = H• + CH3C•FOH -113.5 ±0.2 52.1 -68. ±0.2	97.6±0.2	±0.4
<u>CH3CHFO-H</u> CH3CHFO-H = H• + CH3CHFO• -113.5 ±0.2 52.1 -58.8±0.2	106.8±0.2	±0.4
<u>H-CHFCHFOH</u> H-CHFCHFOH = H• + • CHFCHFOH -154.6±0.6 52.1 -105.6±0.2	101.1±0.4	±0.8
<u>CH2FC-HFOH</u>		

CH2FC-HFOH = H• + CH2FC• FOH -154.6±0.6 52.1 -104.1 ±0.1	102.6±0.35	±0.7
CH2FCHFO-H CH2FCHFO-H = H• + CH2FCHFO• -154.6±0.6 52.1 -99.1±0.1	107.6±0.35	±0.7
CF2-HCH2OH CF2-HCH2OH = H• + • CF2CH2OH -155.3±0.1 52.1 -102.3±0.2	105.1±0.15	±0.3
CF2HCH-HOH CF2HCHOH-H = H• + CF2HC•OH -155.3±0.1 52.1 -102.4±0.2	105.0±0.15	±0.3
CF2HCH2O-H CF2HCH2O-H = H• + CF2HCH2O• -155.3±0.1 52.1 -98.9±0.2	108.5±0.15	±0.3
H-CH2CF2OH H-CH2CF2OH = H• + •CH2CF2OH -174.5±0.1 52.1 -119.4±0.2	107.2±0.15	±0.3
CH3CF2O-H CH3CF2O-H = H• + CH3CF2O• -174.5±0.1 52.1 -110.6±0.1	116.0±0.1	±0.2

* Supporting Information

Supporting information is available, Cartesian Coordinates; Z-matrixes, vibration frequencies, moments of inertia, the method **of** standard deviation, Optimized Geometries, and C-C and C-O internal rotors potential energy profile for target fluorinated ethanol and their related radicals are included.

5. Conclusion

Calculated thermodynamic properties of 19 mono and di-fluoro ethanols and their related radicals using the ab initio and Global-hybrid meta-GGA density function methods are presented. Isodesmic work reactions are employed for the cancellation of calculation errors. Multiple work reactions are utilized to calculate the standard enthalpy of formation at the Gaussian M06-2X calculation level. Optimized geometries and frequencies are used to determine entropy and heat capacity with M06-2x/6-31+g (d,p) level of calculation. Intermolecular torsion potential curves at the M-06-2x/6-31+g (d,p) level of calculation are used to calculate hindered internal rotation contributions to heat capacity and entropy with a correction to the calculated heat capacity and entropy. The calculated Thermochemical properties: Entropy, Heat Capacities at (298 – 1500K), Standard Enthalpy of formation (298K), and the C—F and C—H Bond Dissociation Energies (BDEs) for Mono and Difluorinated Ethanols and Radicals: CH3–xCHFxOH, CH3CH2-xFxOH are presented. The C-H bond energies range from 102.2 to 107.2 Kcal mol^{-1} on the methyl carbons and from 97.3 to 105.2 Kcal mol^{-1} on the secondary ethyl carbons. The calculated values for C-H bond energies for fluorinated methyl carbons are higher than those of the fluorinated ethyl carbons. Calculated values for the O-H bond energies for 2-fluoroethanol are higher than those of O-H bond energies for 1-fluoroethanol and an intermediate calculated value for O-H bond energies for 1,2-difluoroethanol. Introducing a fluorine atom to either methyl or ethyl carbon increases O-H bond energy.

Supplemental Information

Contents:

Table 1. Cartesian Coordinates in Angstroms for target fluorinated ethanol and their related radical's geometries at the m062x/6-31+g (d,p) level of theory

CH2FCH2OH

Center Number	Atomic Number	Cartesian Coordinates X	Y	Z
1	6	0.000003116	-0.000001518	0.000020526
2	9	0.000014524	-0.000002341	-0.000029060
3	1	-0.000005137	-0.000003977	0.000002345
4	1	0.000001564	0.000000767	0.000016765
5	6	-0.000021746	0.000030833	0.000003239
6	1	-0.000002396	0.000010185	0.000006184
7	1	0.000005782	-0.000006013	-0.000000243
8	8	0.000005060	-0.000028332	-0.000029361
9	1	-0.000000767	0.000000396	0.000009605

CH3CHFOH

Center Number	Atomic Number	Cartesian Coordinates X	Y	Z
1	6	0.000021353	0.000006161	-0.000053662
2	6	-0.000117652	0.000027420	0.000072872
3	1	-0.000017542	-0.000011156	-0.000000420
4	1	0.000002761	-0.000007600	0.000015768
5	1	-0.000013646	-0.000002428	0.000033080
6	9	0.000109835	0.000008063	-0.000042913
7	1	0.000018820	-0.000022248	0.000004569
8	8	0.000000477	0.000011019	-0.000059799
9	1	-0.000004406	-0.000009231	0.000030505

C•HFCH2OH

Center Number	Atomic Number	Cartesian Coordinates X	Y	Z
1	6	-0.000020602	0.000003182	-0.000077408
2	9	0.000008151	0.000053880	0.000000565
3	1	-0.000000850	-0.000011335	0.000020759
4	6	0.000027899	-0.000023254	0.000011757

5	1	-0.000003824	-0.000056264	-0.000008549
6	1	0.000008293	0.000001847	-0.000024269
7	8	-0.000008585	-0.000003122	0.000024031
8	1	-0.000010483	0.000035066	0.000053114

CH2FCH•OH

```
------------------------------------------------------------
---------------------------------- Center   Atomic      Cartesian Coordinates
Number   Number     X      Y      Z
------------------------------------------------------------

-------------------------------------------
```

1	6	0.000001249	-0.000000923	-0.000001559
2	9	0.000007760	0.000016425	-0.000015375
3	1	-0.000003551	0.000003238	-0.000020167
4	1	-0.000005664	0.000000417	-0.000013020
5	6	-0.000037395	0.000046885	0.000022542
6	1	0.000001277	-0.000023053	0.000004842
7	8	0.000037801	-0.000035478	-0.000015526
8	1	-0.000001476	-0.000007512	0.000038263

CH2FCH2O•

```
------------------------------------------------------------
---------------------------------- Center   Atomic      Cartesian Coordinates
Number   Number     X      Y      Z
------------------------------------------------------------

-------------------------------------------
```

1	6	0.000009752	0.000030682	0.000120978
2	9	-0.000007540	0.000018428	-0.000104164
3	1	-0.000006175	-0.000009977	-0.000011573
4	1	-0.000001921	-0.000005002	-0.000038255
5	6	-0.000095473	-0.000015917	0.000025447
6	1	0.000036312	-0.000025722	0.000008467
7	1	0.000030627	0.000027608	-0.000011892
8	8	0.000034418	-0.000020100	0.000010992

CH2•CHFOH

```
------------------------------------------------------------
---------------------------------- Center   Atomic      Cartesian Coordinates
Number   Number     X      Y      Z
------------------------------------------------------------

-------------------------------------------
```

1	6	0.000014779	0.000033223	-0.000005569
2	6	-0.000030495	0.000005099	0.000010101
3	1	0.000010354	-0.000009796	-0.000011520
4	1	-0.000018812	-0.000018702	-0.000000754
5	9	0.000041121	0.000009214	-0.000011106

6	1	0.000011136	-0.000017220	0.000008038
7	8	-0.000010627	0.000005847	0.000010223
8	1	-0.000017456	-0.000007665	0.000000588

CH3CF•OH

```
-------------------------------------------------------------------------------
-------------------------------------- Center   Atomic        Cartesian Coordinates
Number  Number    X      Y      Z
-------------------------------------------------------------------------------
------------------------------------------------
```

1	6	0.000001336	0.000012708	-0.000046895
2	6	-0.000071306	0.000036407	0.000029101
3	1	-0.000014952	-0.000021374	-0.000009157
4	1	-0.000000203	0.000002487	0.000002208
5	1	-0.000029193	0.000004057	0.000022548
6	9	0.000068515	0.000006342	-0.000002103
7	8	0.000028358	-0.000032227	0.000010729
8	1	0.000017444	-0.000008399	-0.000006430

CH3CHFO•

```
-------------------------------------------------------------------------------
-------------------------------------- Center   Atomic        Cartesian Coordinates
Number  Number    X      Y      Z
-------------------------------------------------------------------------------
------------------------------------------------
```

1	6	-0.000025583	0.000026726	0.000006931
2	6	-0.000004925	-0.000017774	-0.000022873
3	1	-0.000016576	0.000000727	-0.000013731
4	1	-0.000010431	0.000004336	0.000014687
5	1	-0.000006608	0.000007692	-0.000003229
6	9	0.000038154	-0.000008832	0.000013405
7	1	0.000017550	-0.000008997	-0.000000815
8	8	0.000008419	-0.000003878	0.000005626

CH2FCHFOH

```
-------------------------------------------------------------------------------
-----------------------------------------------
Center   Atomic        Cartesian Coordinates
Number  Number    X      Y      Z
-------------------------------------------------------------------------------
------------------------------------------------
```

1	6	-0.000209387	-0.000014742	-0.000046791
2	6	-0.000078791	-0.000026117	-0.000054220
3	9	0.000055970	-0.000014001	0.000038779
4	1	0.000037582	0.000003431	0.000025080

5	1	0.000008069	-0.000003401	0.000000786
6	9	0.000213499	-0.000011196	0.000078222
7	1	-0.000002897	0.000018025	-0.000024730
8	8	-0.000003076	0.000003947	-0.000028849
9	1	-0.000020969	0.000044055	0.000011724

CF2HCH2OH

| Center | Atomic | Cartesian Coordinates | | |
| Number | Number | X | Y | Z |

1	6	-0.000006264	-0.000010001	-0.000015081
2	6	0.000196771	-0.000004891	0.000062245
3	9	-0.000177152	-0.000024722	-0.000031629
4	9	0.000058924	0.000072953	-0.000005858
5	1	-0.000009820	0.000002957	-0.000026980
6	1	-0.000016810	-0.000024610	0.000029199
7	1	-0.000014380	0.000016248	-0.000024677
8	8	-0.000035113	-0.000034435	0.000010637
9	1	0.000003845	0.000006502	0.000002145

CH3CF2OH

| Center | Atomic | Cartesian Coordinates | | |
| Number | Number | X | Y | Z |

1	6	0.000056837	-0.000094451	0.000110833
2	6	-0.000057651	0.000098699	0.000001478
3	1	0.000011317	-0.000004093	-0.000026345
4	1	0.000014600	-0.000025891	-0.000015228
5	1	-0.000001272	-0.000012112	-0.000025019
6	9	-0.000131047	-0.000088396	0.000020206
7	9	0.000140461	0.000063879	0.000020905
8	8	-0.000014132	0.000020314	-0.000026071
9	1	-0.000019113	0.000042051	-0.000060759

CH2FC•FOH

| Center | Atomic | Cartesian Coordinates |

---   1   6   -0.000085873 -0.000008147 -
0.000033316
    2   6   0.000107121 -0.000054590 0.000053325
    3   9   -0.000086887 -0.000026092 -0.000005448
    4   1   -0.000015313 -0.000013867 -0.000003160
    5   1   0.000000240 -0.000001842 -0.000016239
    6   9   0.000003934 0.000030284 0.000010058
    7   8   0.000076030 0.000051318 -0.000056283
    8   1   0.000000747 0.000022936 0.000051063

C•HFCFHOH

Center Atomic   Cartesian Coordinates
Number Number   X   Y   Z

----

    1   6   -0.000058756 0.000023216 0.000035904
    2   6   -0.000088748 -0.000108788 -0.000099552
    3   9   0.000068251 0.000068970 0.000065027
    4   1   0.000023870 0.000025841 0.000034527
    5   9   0.000011825 -0.000021999 -0.000019250
    6   1   -0.000011611 0.000023111 -0.000006306
    7   8   0.000047202 0.000013830 -0.000008891
    8   1   0.000007966 -0.000024181 -0.000001459

CH2FCHFO•

-- ---

Center Atomic   Cartesian Coordinates
Number Number   X   Y   Z

----

    1   6   -0.000043639 0.000033246 0.000009821
    2   6   0.000048716 -0.000026830 -0.000018036
    3   9   -0.000022444 0.000006378 -0.000008784
    4   1   0.000001695 0.000020117 0.000006913
    5   1   0.000003424 0.000009684 0.000001110
    6   9   0.000023881 0.000019171 0.000030292
    7   1   0.000006948 -0.000005119 0.000004871
    8   8   -0.000018580 -0.000056646 -0.000026187

CHF2C•HOH

Center   Atomic   Cartesian Coordinates
Number   Number   X   Y   Z

----
   1   6   0.000004676 -0.000070251   0.000051141
   2   6   0.000027275 -0.000132737 -0.000118944
   3   1   -0.000013476   0.000008382   0.000015950
   4   8   0.000006557   0.000005715 -0.000023257
   5   1   0.000018970 -0.000003557   0.000007094
   6   1   -0.000051957 -0.000006265 -0.000015262
   7   9   -0.000013983   0.000033076   0.000051050
   8   9   0.000021937   0.000165638   0.000032227

CF2•CFOH

Center   Atomic   Cartesian Coordinates
Number   Number   X   Y   Z
- ---
----
   1   6   0.000045030   0.000052085 -0.000051151
   2   6   0.000001125   0.000036125   0.000024114
   3   1   -0.000002716   0.000001155   0.000006275
   4   8   -0.000050429 -0.000036049   0.000025853
   5   1   -0.000005831 -0.000009141 -0.000009552
   6   1   -0.000015223   0.000001185   0.000012909
   7   9   0.000011981 -0.000006184   0.000002242
   8   9   0.000016063 -0.000039174 -0.000010691

CHF2CH2O•

-- Center   Atomic   Cartesian Coordinates
Number   Number   X   Y   Z

----
   1   6   -0.000032920 -0.000119205   0.000076072
   2   6   -0.000030593 -0.000022551 -0.000053428
   3   9   -0.000036099   0.000010735   0.000025263
   4   9   -0.000006345 -0.000017725 -0.000018129
   5   1   -0.000014790 -0.000026477   0.000021461
   6   1   0.000040237   0.000019571 -0.000021746
   7   1   0.000038791   0.000046905 -0.000010985
   8   8   0.000041718   0.000108747 -0.000018508

CH2•CF2OH

```
------------------------------------------------------------------------------------------------

-----------------------------------------------

Center  Atomic      Cartesian Coordinates
Number  Number    X      Y      Z
------------------------------------------------------------------------------------------------

-------------------------------------------------      1   6    0.000009221 0.000001893 -
0.000000440
  2   6   -0.000001071 -0.000001757 -0.000000838
  3   1    0.000004590 0.000002074 0.000000801
  4   1    0.000019461 0.000007677 -0.000000682
  5   9   -0.000009537 -0.000013694 -0.000009305
  6   9   -0.000017948 0.000004296 0.000010263
  7   8    0.000001500 0.000001265 0.000000690
  8   1   -0.000006214 -0.000001754 -0.000000488
```

CH3CF2O•

```
------------------------------------------------------------------------------------------------

---------------------------------------------

Center  Atomic      Cartesian Coordinates
Number  Number    X      Y      Z
------------------------------------------------------------------------------------------------

-----------------------------------------------

  1   6   -0.000049808 0.000049564 0.000101113
  2   6    0.000314370 -0.000446227 -0.000322008
  3   1   -0.000001882 0.000003042 0.000019595
  4   1    0.000007442 0.000003265 -0.000034062
  5   1    0.000007479 -0.000020681 0.000033341
  6   9   -0.000225548 0.000114125 0.000063416
  7   9   -0.000008257 0.000232988 0.000078418
  8   8   -0.000043796 0.000063924 0.000060187
```

Table 2. Frequencies for target fluorinated ethanol and their related radicals at the m062x/6-31+g (d,p) level of theory (cm-1)

CH2FCH2OH	153.7115	318.4318	395.2338
	526.9463	879.0892	913.6087
	1084.2017	1123.0654	1155.2746
	1227.6058	1278.1133	1382.4638
	1406.9769	1443.0414	1499.5695
	1503.3220	3040.8126	3082.2806
	3119.3826	3144.6188	3895.3665
CH3CHFOH	222.4821	381.0906	395.9003
	476.6613	591.2739	882.7961

	967.7343	1083.1639	1139.5797
	1207.0636	1299.2263	1395.3640
	1408.6162	1481.8193	1491.1530
	1496.5668	3074.8833	3098.4238
	3162.1117	3168.6137	3905.4122
C•HFCH2OH	117.2311	305.3799	387.2015
	487.0844	672.8813	897.2987
	1047.2118	1078.4074	1202.2847
	1230.9543	1340.0387	1397.8007
	1433.9762	1486.8167	3060.3820
	3135.9012	3227.9702	3904.1459
CH2FCH•OH	136.3261	293.7121	372.9921
	469.7867	588.3163	948.2198
	976.5181	1064.6622	1237.5557
	1261.6997	1324.3255	1380.0377
	1477.7581	1517.2468	3089.0157
	3148.4826	3209.7826	3921.1700
CH2FCH2O•	138.8341	326.2587	461.7143
	645.0346	900.2694	995.0829
	095.1447	1132.5555	1209.7403
	287.6953	1366.6408	1397.9225
	436.1160	1499.8795	2978.3111
	021.0190	3079.4857	3141.2839
CH2•CHFOH	103.9492	354.4677	378.6019
	464.4234	533.4936	615.0872
	917.9447	973.1708	1082.8162
	1204.3179	1292.9244	1375.2909
	1421.9543	1501.7259	3093.4946
	3176.3175	3301.9307	3902.4787
CH3CF•OH	183.0404	339.6091	381.8885
	446.3670	566.0359	874.2810
	992.1200	1063.5695	1140.5736
	1302.2104	1363.7386	1442.6873
	1477.4645	1485.2396	3031.7545
	3128.3197	3174.6193	3874.3729
CH3CHFO•	234.8451	359.2559	443.1211
	572.8867	866.7266	929.6382
	1040.2264	1088.1878	1158.3501
	1216.4848	1326.7171	1391.8028
	1484.8041	1492.1170	2997.0767
	3077.4025	3170.0581	3179.9978
CH2FCHFOH	123.7297	253.3055	341.9812
	442.3530	508.5501	585.7551
	932.9529	1049.0131	1096.9489
	1132.3163	1216.5809	1244.5656
	1341.7047	1356.5333	1432.8601
	1477.0191	1509.6783	3073.3252
	3141.2603	3156.9890	3868.8911
CF2HCH2OH	261.5051	333.1831	429.7499
	803.8978	905.9089	1069.3307
	1113.2025	1153.4006	1280.2493

	1377.1368	1403.4215	1433.7765
	1493.5768	1500.2248	1525.4172
	3032.2176	3044.0073	3104.6932
	3125.0745	3137.9412	3910.0283
CH3CF2OH	213.0609	336.6797	378.8026
	407.4641	552.4402	575.8400
	584.5503	847.6087	959.8589
	1000.3581	1135.7264	1204.1092
	1301.5094	1390.3411	1483.7215
	1488.5073	1492.2192	3086.0281
	3175.9594	3186.0332	3866.1356
CH2FC•FOH	109.4115	244.9232	388.6686
	436.4869	472.3557	575.6453
	917.0734	1055.3780	1102.3841
	1135.9138	1244.9454	1336.2661
	1365.0202	1448.3398	1502.9979
	3049.1641	3148.5710	3871.9507
C•HFCFHOH	107.9351	257.2006	353.4210
	416.6962	546.9238	646.1233
	765.4796	916.4141	991.0405
	1161.1762	1227.5285	1327.6897
	1356.7224	1403.4567	1473.0076
	3148.9051	3254.0619	3863.4809
CH2FCHFO•	122.5173	253.1428	406.6622
	426.8055	595.6816	922.4453
	981.5646	1062.7404	1130.5318
	1159.6593	1193.8858	1264.5276
	1306.1655	1411.2155	1509.8582
	3020.9800	3106.4335	3176.7438
CHF2C•HOH	102.5287	255.5062	307.6957
	498.7454	531.8669	588.6407
	776.4400	943.7162	1040.6628
	1117.9847	1203.9780	1346.8684
	1361.3396	1407.7937	1499.6505
	3165.5275	3252.1304	3846.9473
C•F2CH2OH	111.2059	240.3480	355.7645
	439.8572	476.0825	590.2258
	889.8898	1035.2006	1136.6153
	1208.8861	1259.6672	1294.5087
	1380.7274	1441.8811	1496.7631
	3013.5503	3147.9499	3898.0279
CHF2CH2O•	128.4545	257.5539	433.4387
	494.0758	586.0455	917.5567
	1096.8914	1139.6268	1157.8405
	1232.6325	1307.2674	1385.1305
	1404.8759	1510.7673	1859.0406
	3031.4094	3120.2437	3174.3851
CH2•CF2OH	90.5157	313.8094	380.5941
	392.3412	492.4899	547.1101
	593.9731	597.4030	869.8795
	960.9120	1070.4557	1131.1978

	1308.1224	1403.4875	1499.7334
	3211.2694	3342.0724	3863.9788
CH3CF2O•	168.6127	340.9654	360.5945
	534.8395	551.7060	557.0587
	846.1077	915.5633	974.0145
	1162.5612	1220.2174	1281.3277
	1397.7811	1466.2734	1488.6608
	3077.0770	3170.5354	3190.3722

Table 3. Moment of inertia for target fluorinated ethanol and their related radicals at the m062x/6-31+g (d,p) level of theory (GHZ)

CH2FCH2OH	15.7788859 5.5493007 4.6093694
CH3CHFOH	10.0766267 8.3371985 5.1680915
C•HFCH2OH	18.77997 5.13190 4.51693
CH2FCH•OH	23.2409534 4.5476319 4.3485670
CH2FCH2O•	18.24036 5.23748 4.59279
CH2•CHFOH	10.3760339 9.1250064 5.3599040
CH3CF•OH	10.20355 9.05876 5.07651
CH3CHFO•	10.7702750 8.5850705 5.3185779
CH2FCHFOH	9.4161023 3.5112856 2.7600267
CF2HCH2OH	34.8864548 9.2467926 8.1783038
CH3CF2OH	9.4161023 3.5112856 2.7600267
CH2FC•FOH	9.7798799 3.5506736 2.7289504
C•HFCFHOH	9.5295647 3.7425748 2.9008390
CH2FCHFO•	10.2331738 3.5038934 2.8019184
CHF2C•HOH	9.1000423 3.8971695 3.0141244
C•F2CH2OH	9.6401202 3.8114853 2.8520087
CHF2CH2O•	9.21258 3.85327 2.93914
CH2•CF2OH	5.7024545 5.2730634 5.1759614

CH3CF2O•	5.9149594 5.2219748 5.0070937

Figure 1. Optimized Geometry for CH2FCH2OH at the m062x/6-31+g(d,p) level of theory

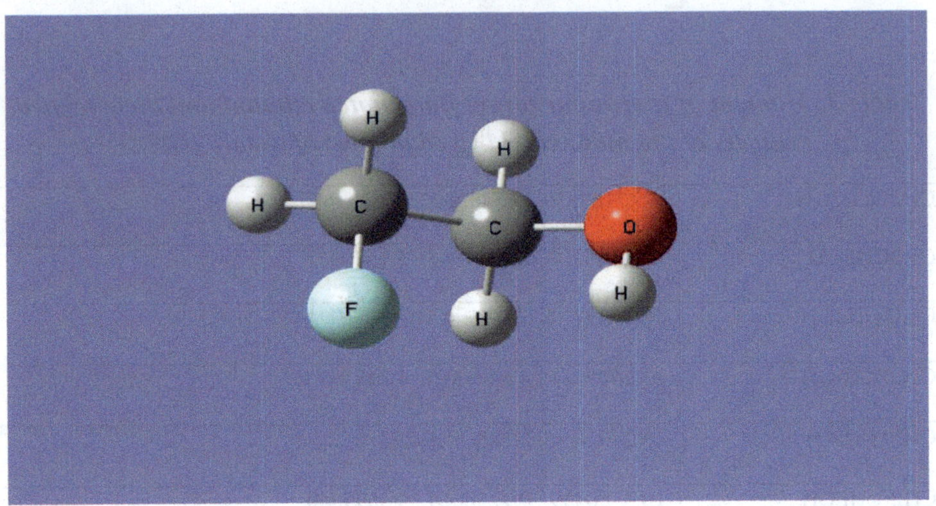

Figure 2. Optimized Geometry for CH3CHFOH at the m062x/6-31+g(d,p) level of theory

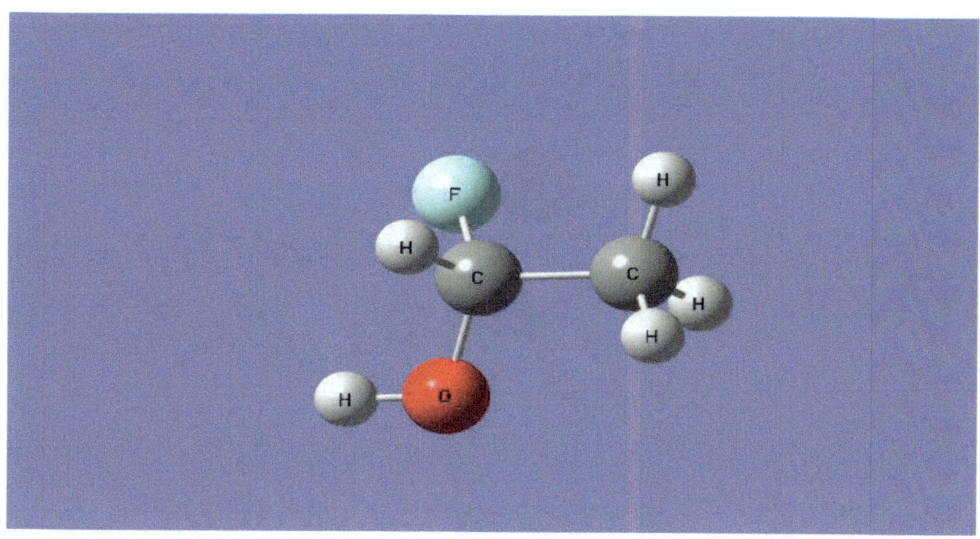

Figure 3. Optimized Geometry for C•HFCH2OH at the m062x/6-31+g(d,p) level of theory

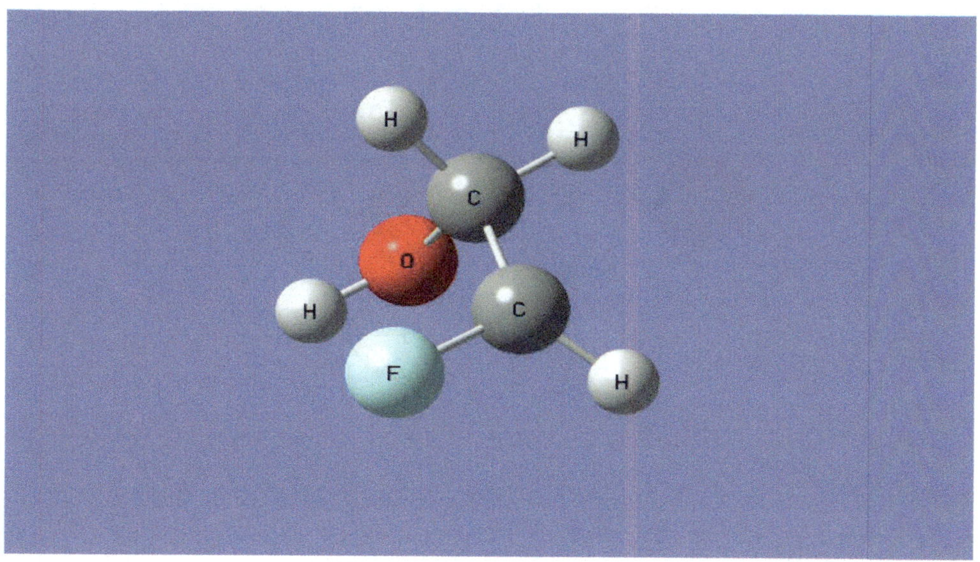

Figure 4. Optimized Geometry for CH2FCH•OH at the m062x/6-31+g(d,p) level of theory

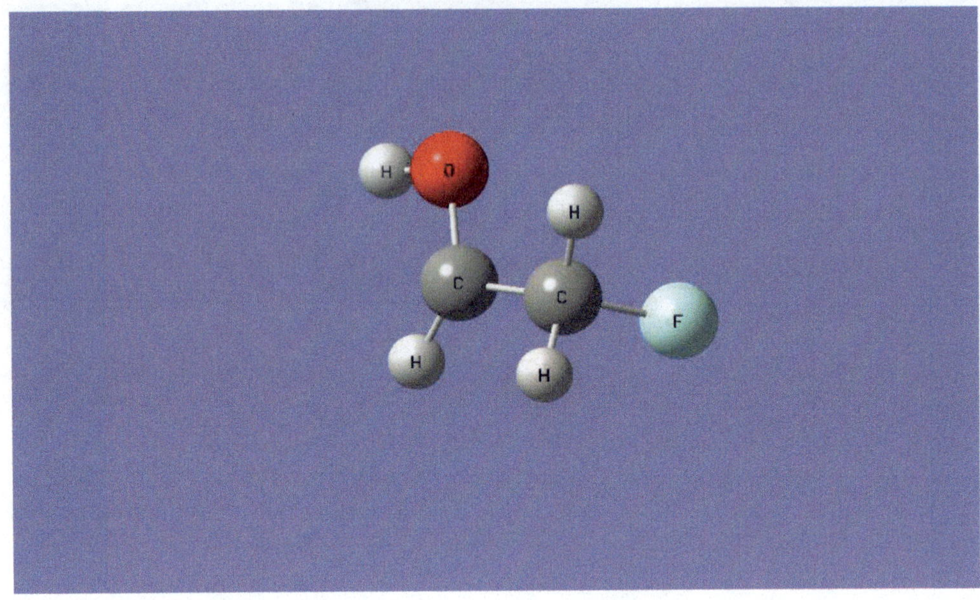

Figure 5. Optimized Geometry for CH2FCH2O• at the m062x/6-31+g(d,p) level of theory

Figure 6. Optimized Geometry for CH2•CHFOH at the m062x/6-31+g(d,p) level of theory

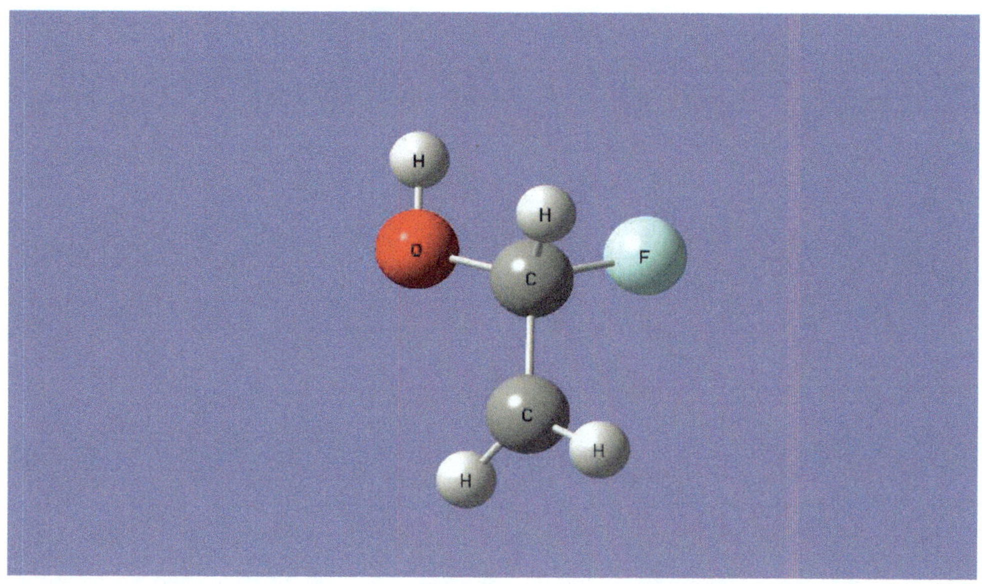

Figure 7. Optimized Geometry for CH3CF•OH at the m062x/6-31+g(d,p) level of theory

Figure 8. Optimized Geometry for CH3CHFO• at the m062x/6-31+g(d,p) level of theory

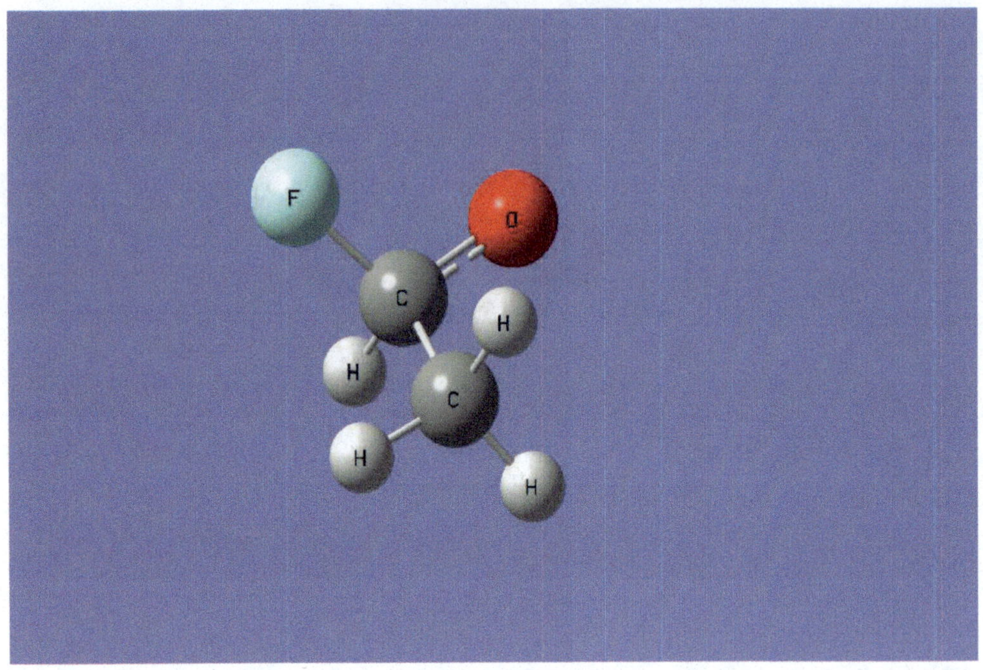

Figure 9. Optimized Geometry for CH2FCHFOH at the m062x/6-31+g(d,p) level of theory

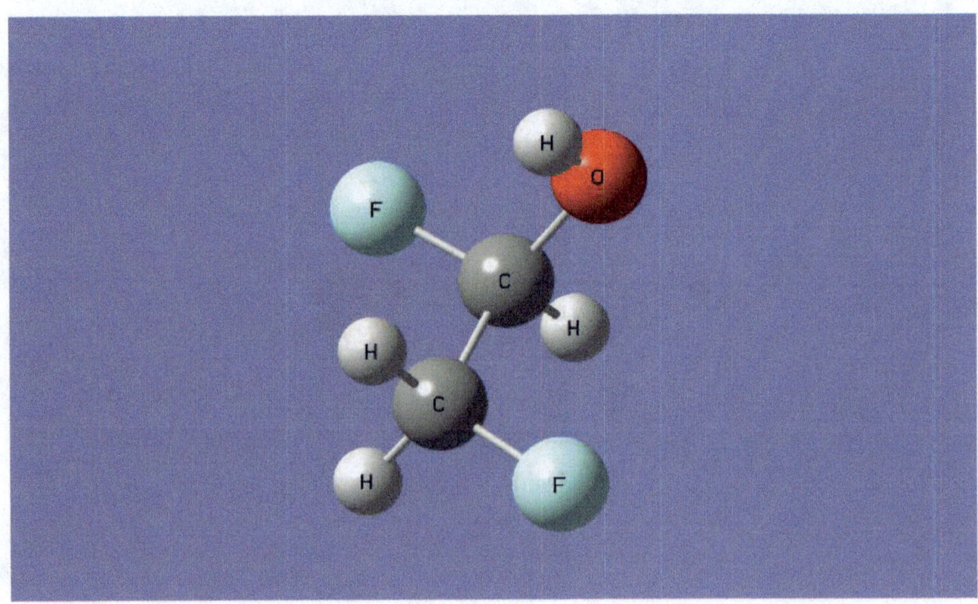

Figure 10. Optimized Geometry for CF2HCH2OH at the m062x/6-31+g(d,p) level of theory

Figure 11. Optimized Geometry for CH3CF2OH at the m062x/6-31+g(d,p) level of theory

Figure 12. Optimized Geometry for C•HFCFHOH at the m062x/6-31+g(d,p) level of theory

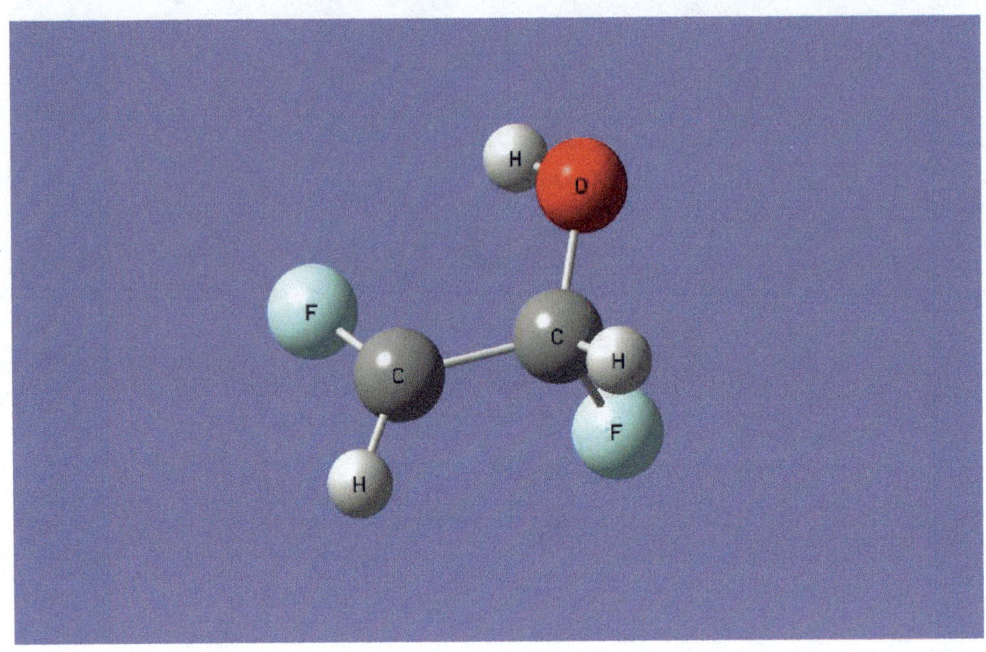

Figure 13. Optimized Geometry for CH2FC•FOH at the m062x/6-31+g(d,p) level of theory

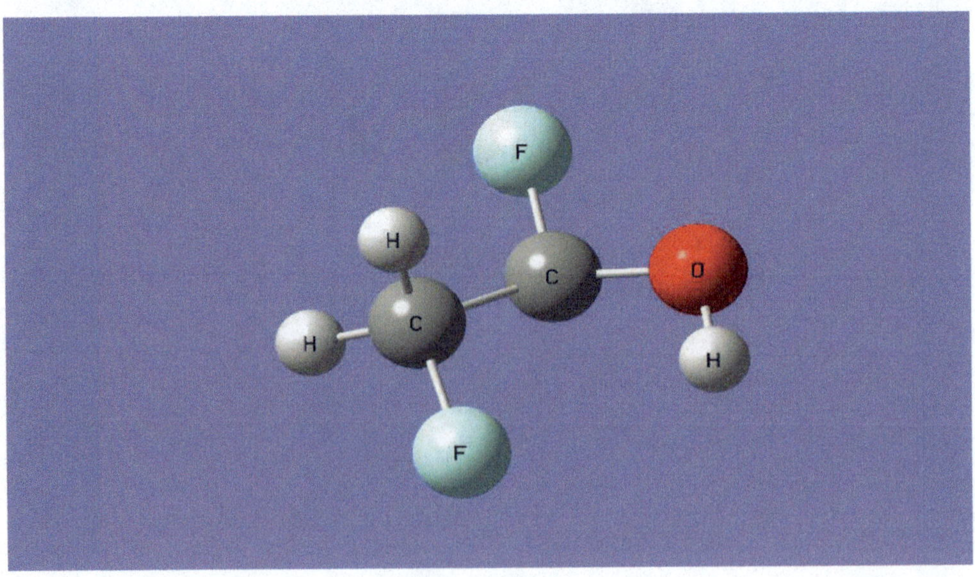

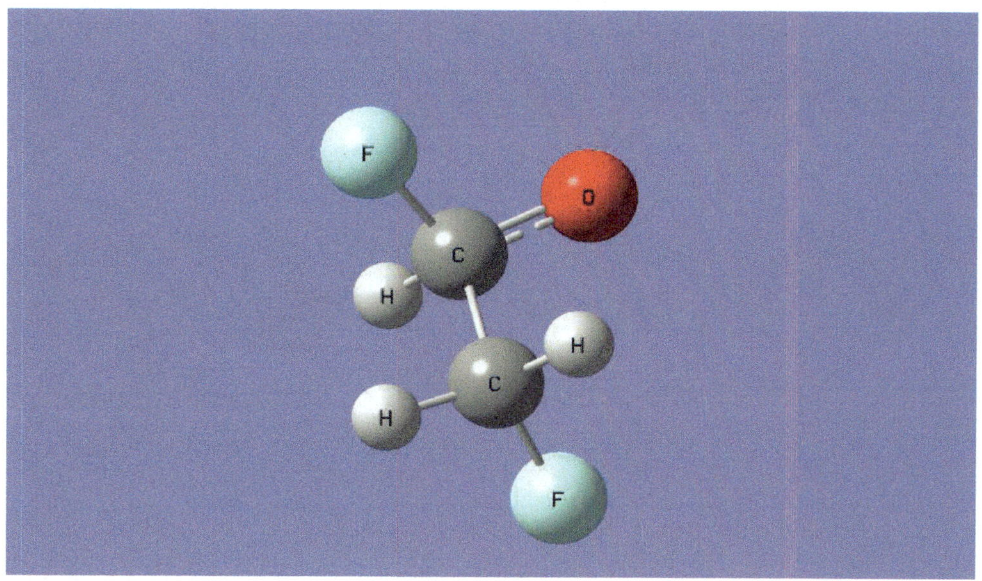

Figure 13. Optimized Geometry for CH2FCHFO• at the m062x/6-31+g(d,p) level of theory

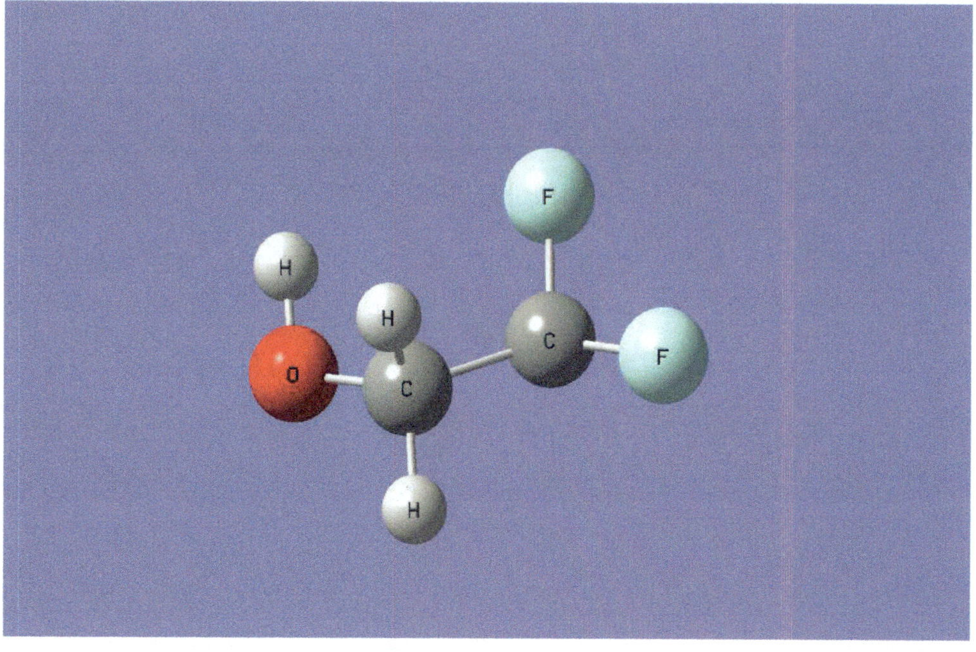

Figure 14. Optimized Geometry for C•F2CH2OH at the m062x/6-31+g(d,p) level of theory

Figure 15. Optimized Geometry for CHF2C•HOH at the m062x/6-31+g(d,p) level of theory

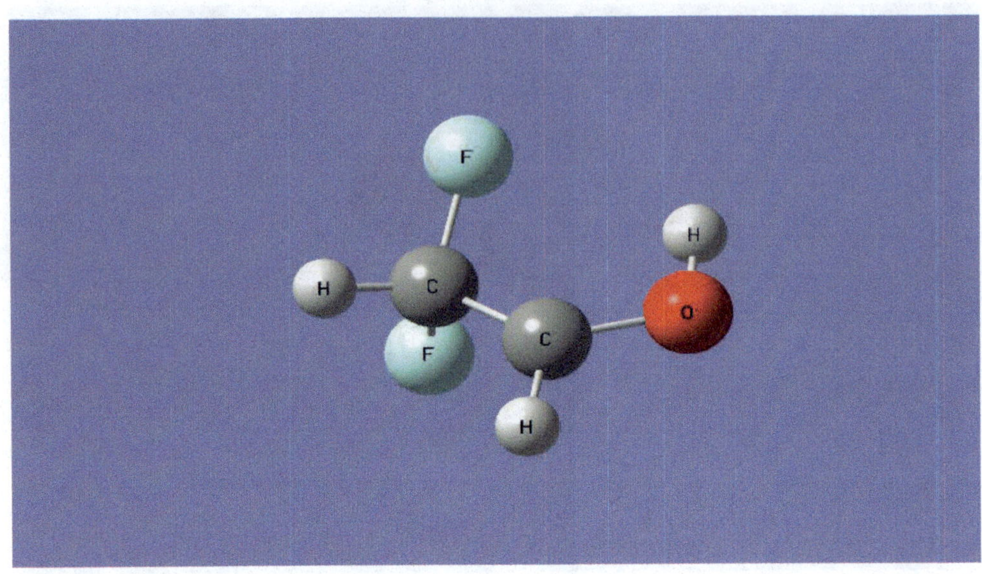

Figure 16. Optimized Geometry for CHF2CH2O• at the m062x/6-31+g(d,p) level of theory

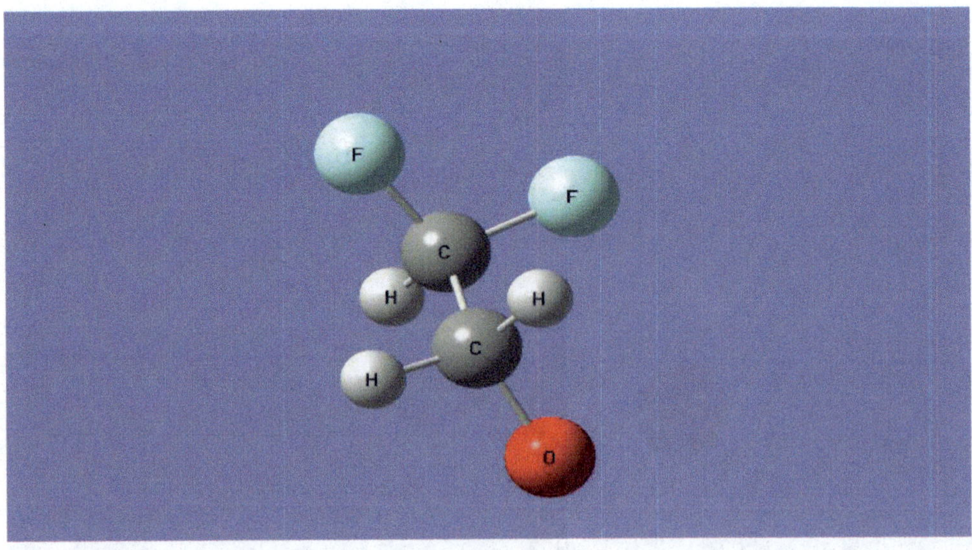

Figure 17. Optimized Geometry for CH2•CF2OH at the m062x/6-31+g(d,p) level of theory

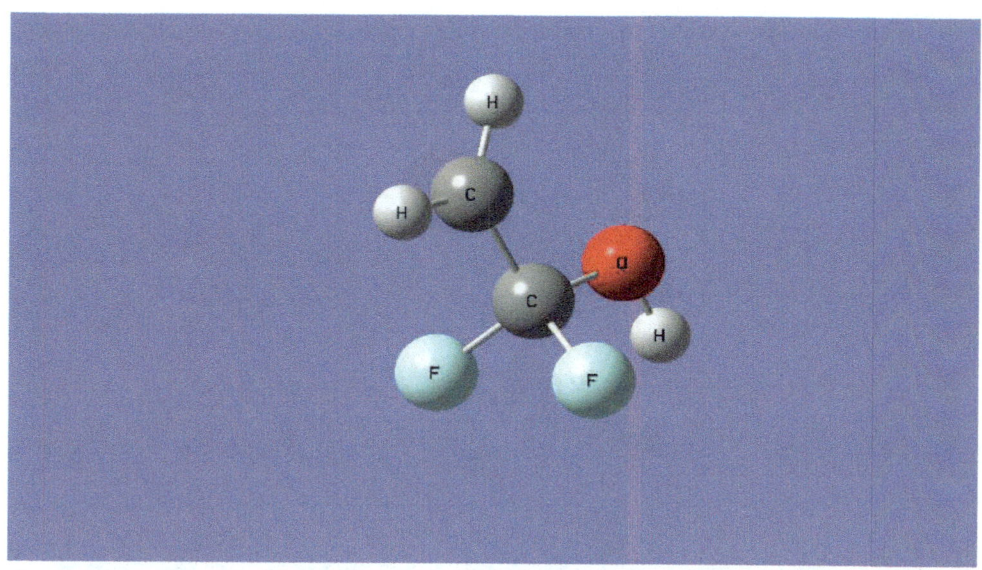

Figure 18. Optimized Geometry for CH2•CF2OH at the m062x/6-31+g(d,p) level of theory

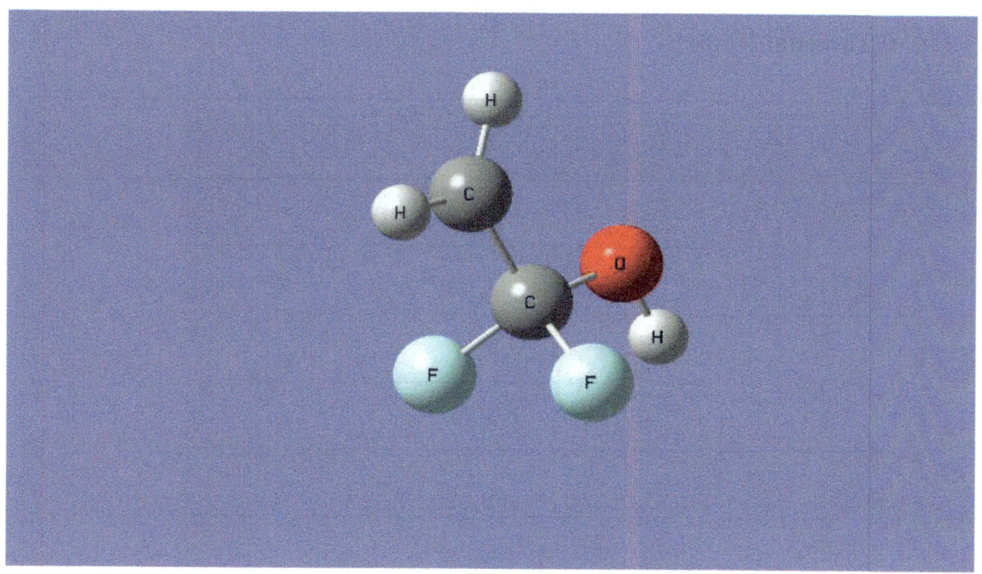

Figure 19. Optimized Geometry for CH3CF2O• at the m062x/6-31+g(d,p) level of theory

Figure 20. Potential energy profile of C-C and C-O internal rotors for CH2FCH2OH. The solid lines indicate Fourier series expansion

C-C Internal Rotor

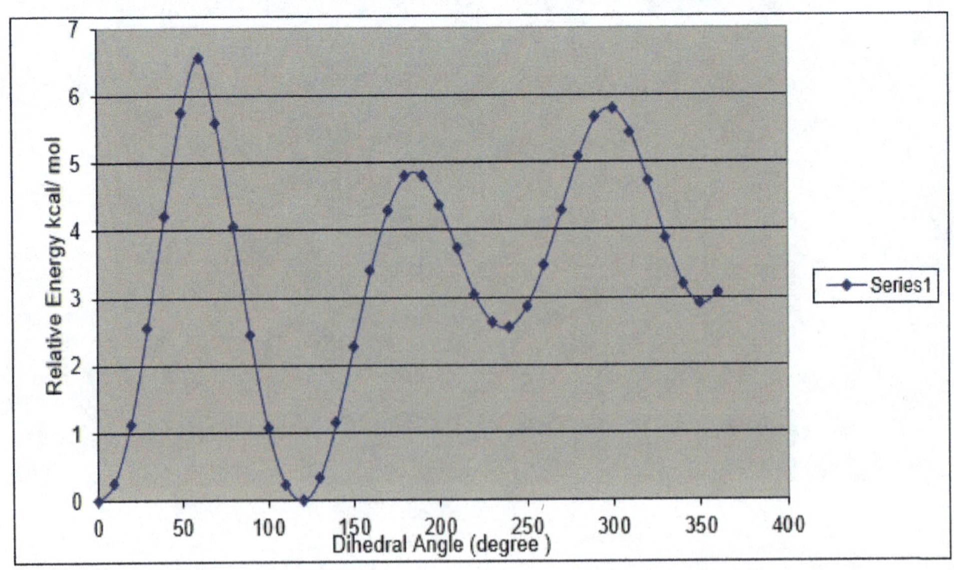

C-O Internal Rotor

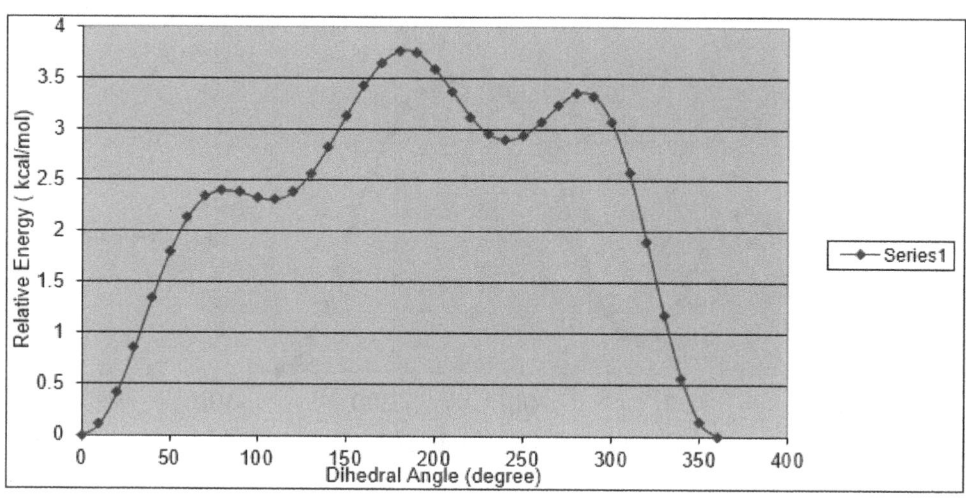

Figure 21. Potential energy profile of C-C and C-O internal rotors for CH3CHFOH. The solid lines indicate Fourier series expansion

C-C Internal Rotor

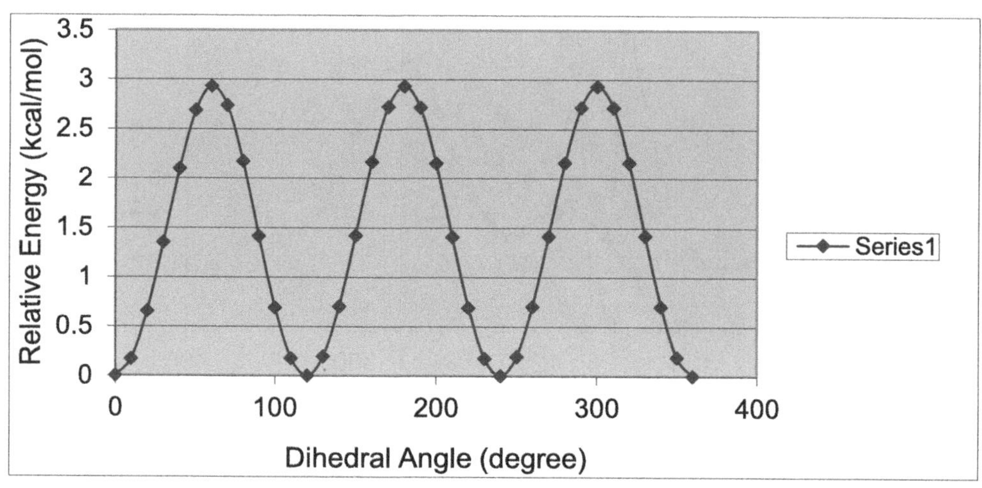

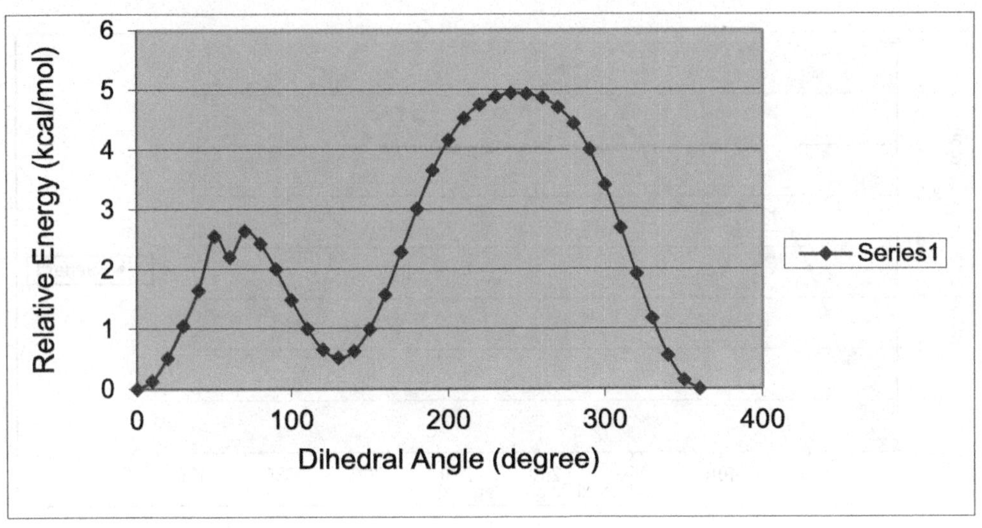

Figure 22. Potential energy profile of C-C and C-O internal rotors for C•HFCH2OH. The solid lines indicate Fourier series expansion

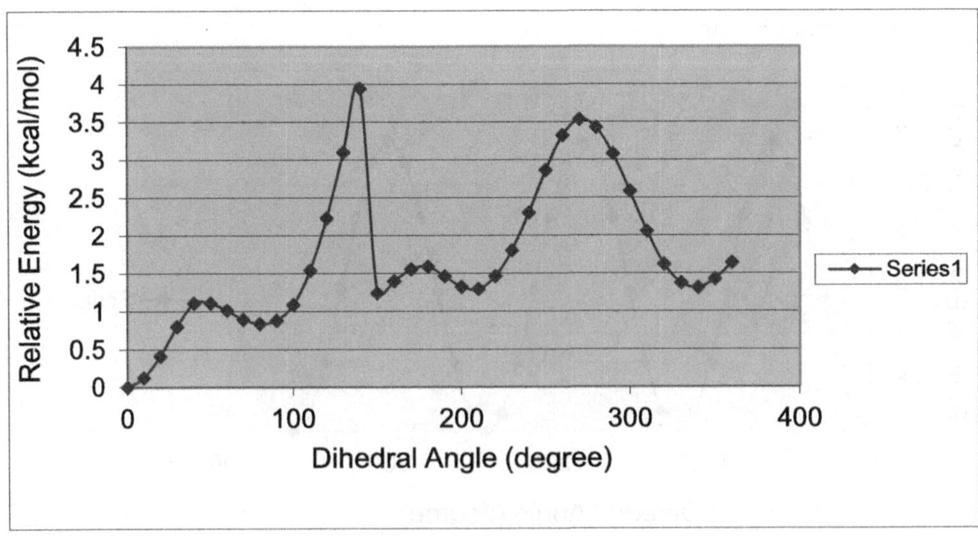

C-O Internal Rotor

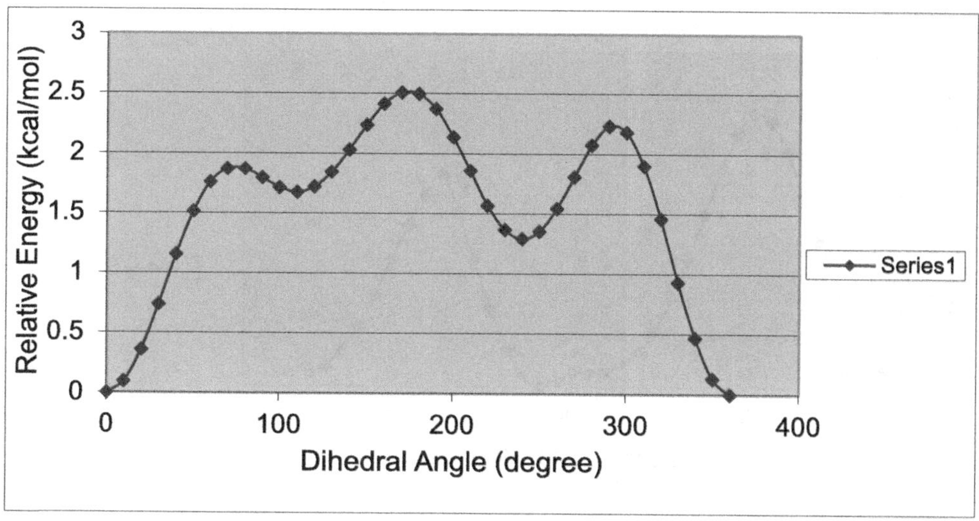

Figure 23. Potential energy profile of C-C and C-O internal rotors for CH2FCH•OH. The solid lines indicate Fourier series expansion

C-C Internal Rotor

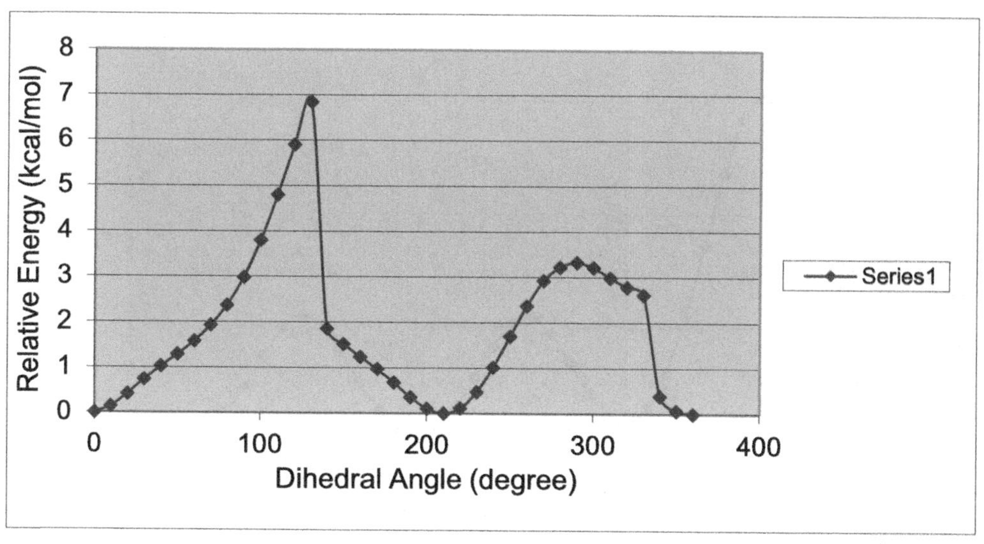

C-O Internal Rotor

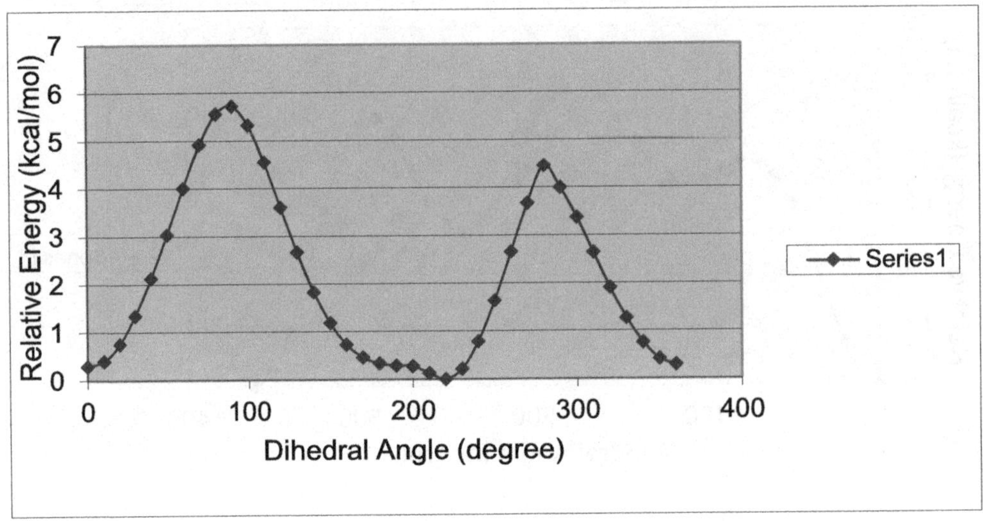

Figure 24. Potential energy profile of C-C and C-O internal rotors for CH2FCH2O•. The solid lines indicate Fourier series expansion

C-C Internal Rotor

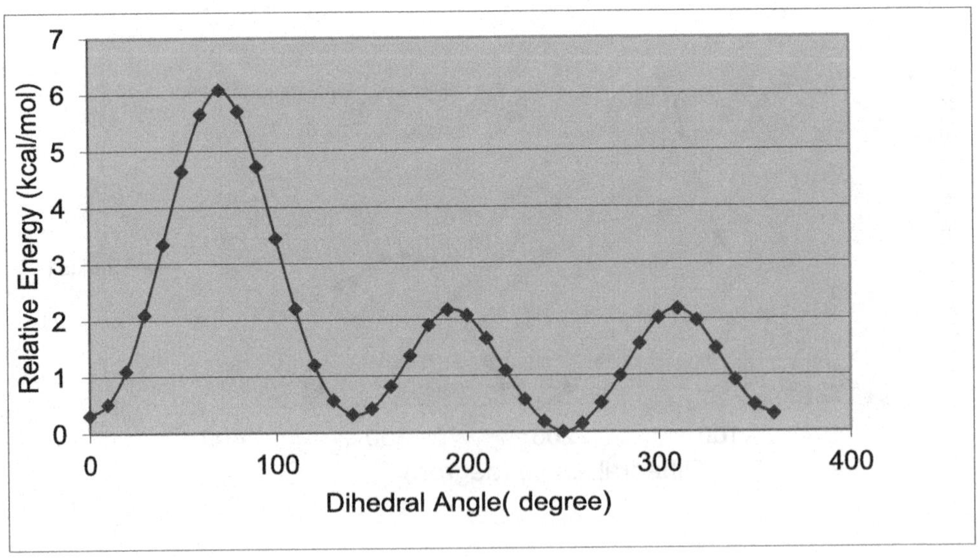

Figure 25. Potential energy profile of C-C and C-O internal rotors for CH2•CHFOH.The solid lines indicate Fourier series expansion

C-C Internal Rotor

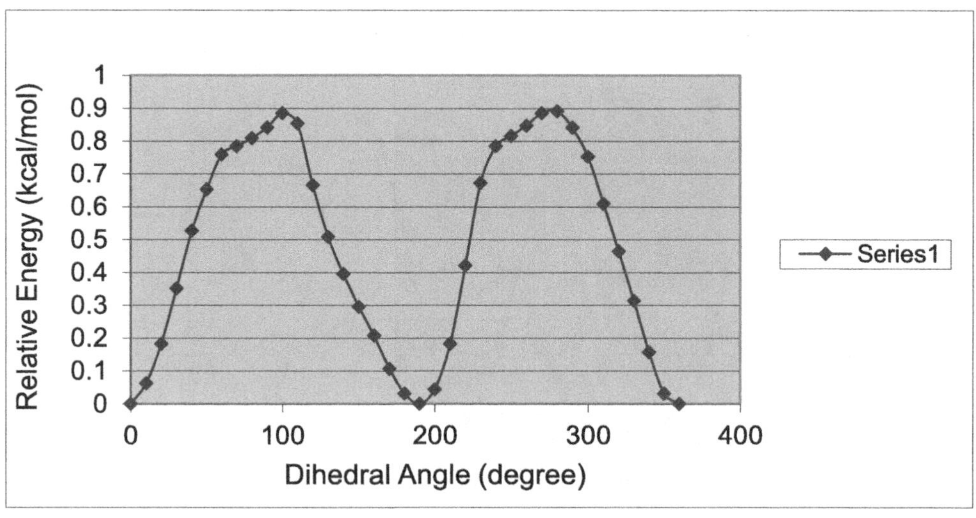

C-O Internal Rotor

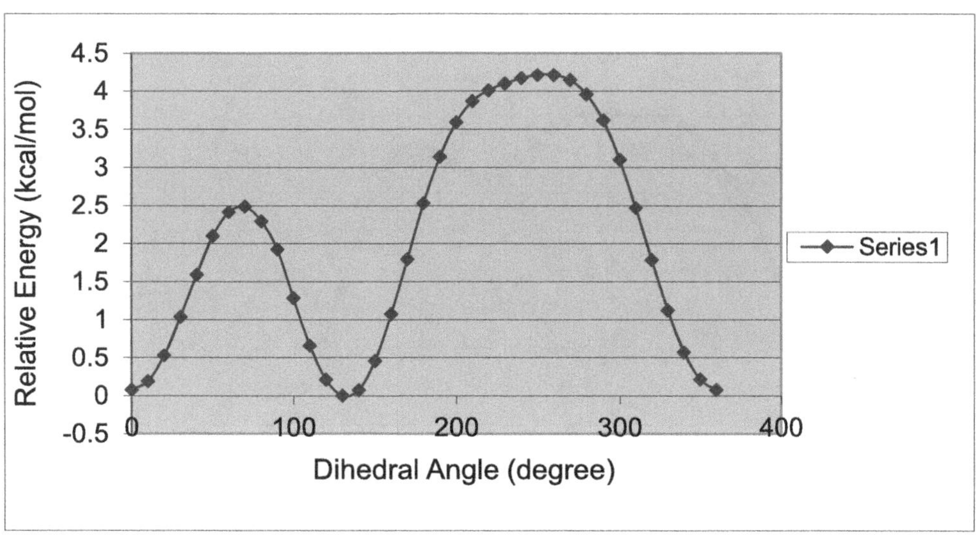

Figure 26. Potential energy profile of C-C and C-O internal rotors for CH3CF•OH.The solid lines indicate Fourier series expansion

C-C Internal Rotor

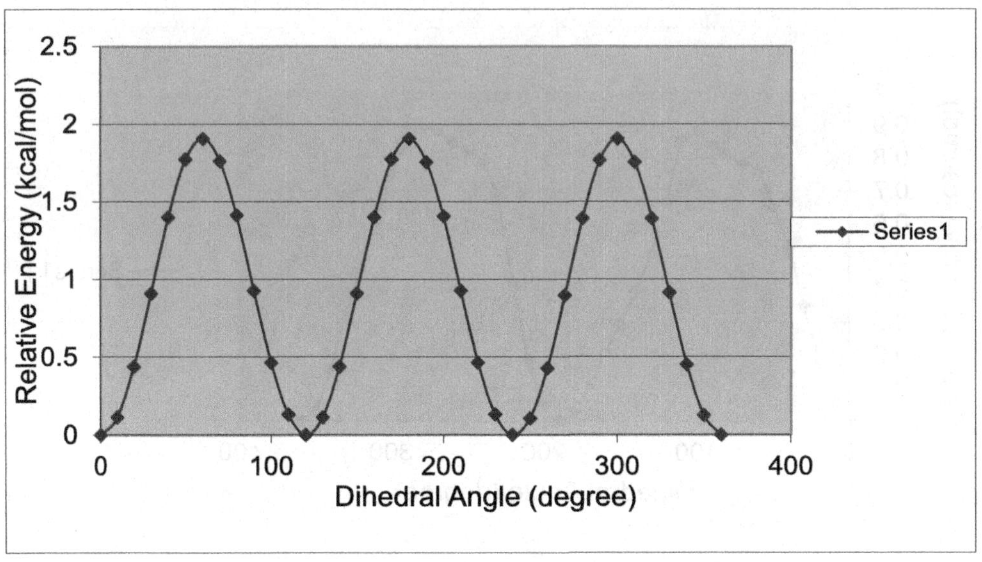

C-O Internal Rotor

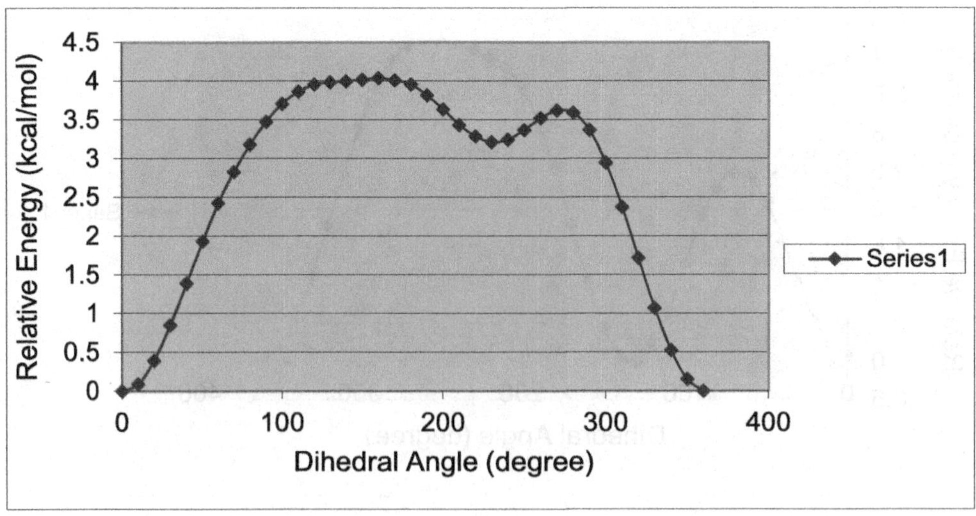

Figure 27. Potential energy profile of C-C and C-O internal rotors for CH3CHFO•. The solid lines indicate Fourier series expansion

C-C Internal Rotor

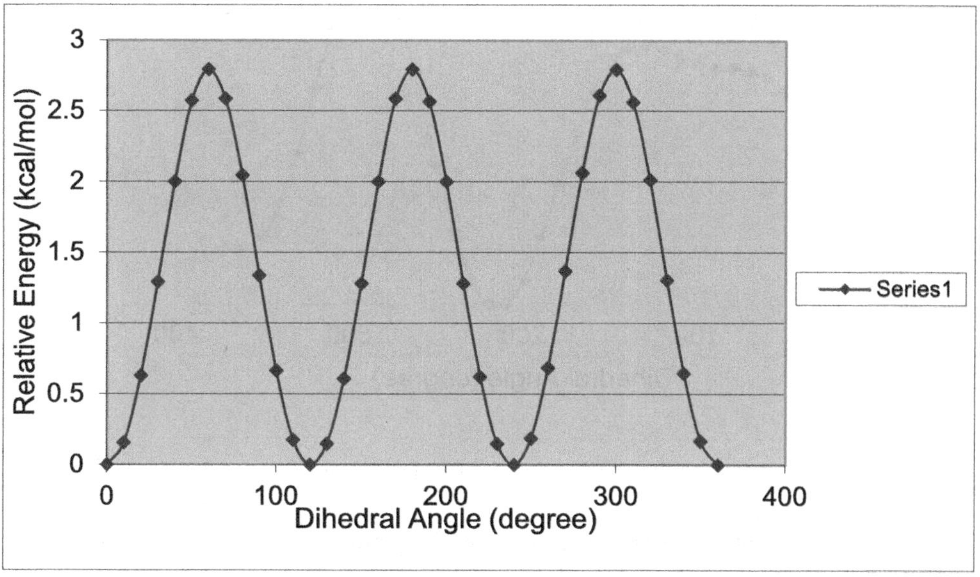

Figure 28. Potential energy profile of C-C and C-O internal rotors for CH2FCHFOH. The solid lines indicate Fourier series expansion

C-C Internal Rotor

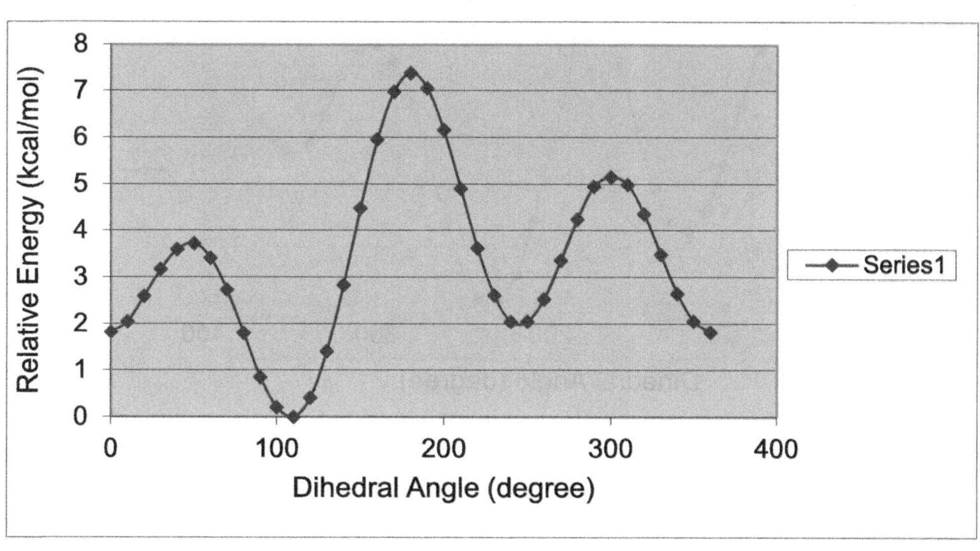

C-O Internal Rotor

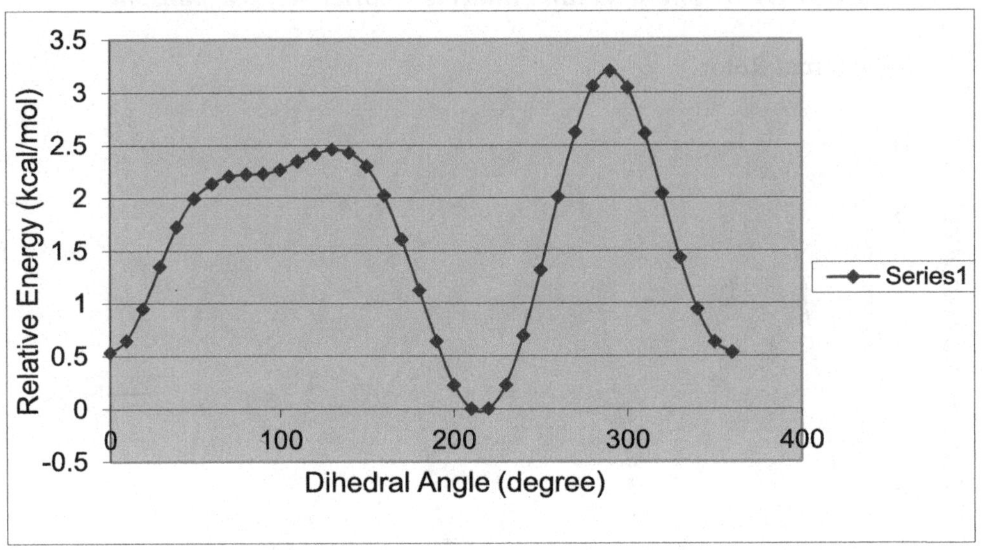

Figure 29. Potential energy profile of C-C and C-O internal rotors for CF2HCH2OH. The solid lines indicate Fourier series expansion

C-C Internal Rotor

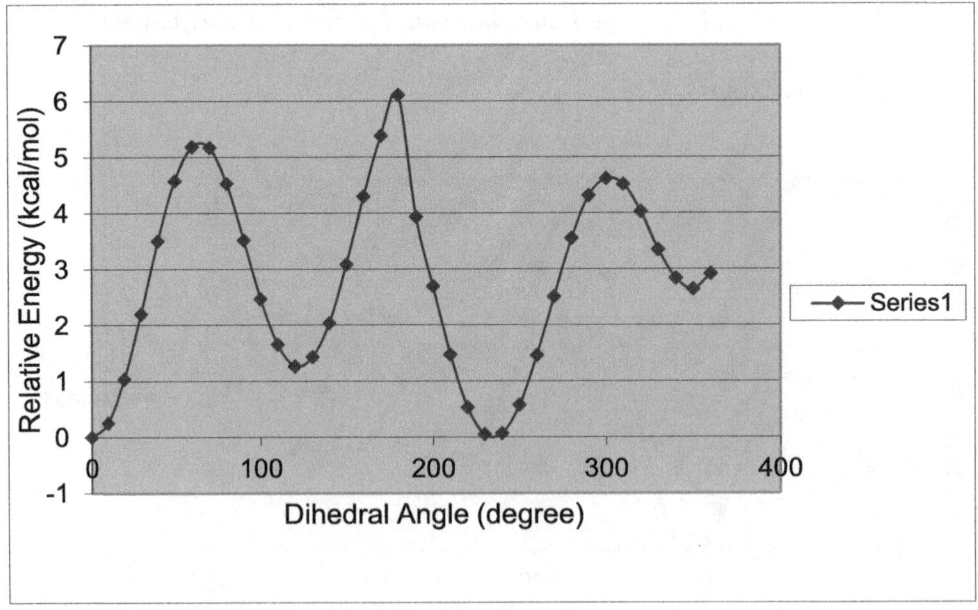

C-O Internal Rotor

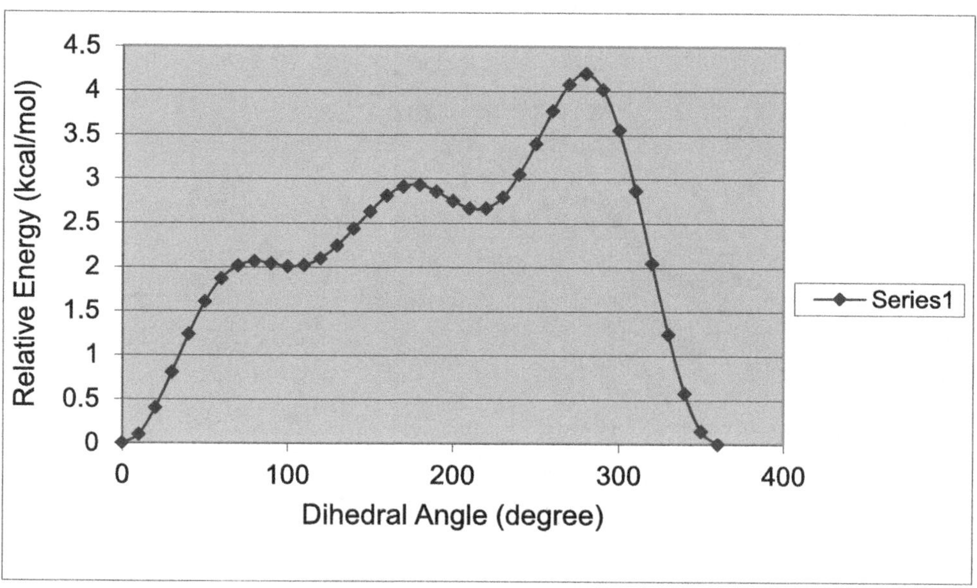

Figure 30. Potential energy profile of C-C and C-O internal rotors for CH3CF2OH. The solid lines indicate Fourier series expansion

C-C Internal Rotor

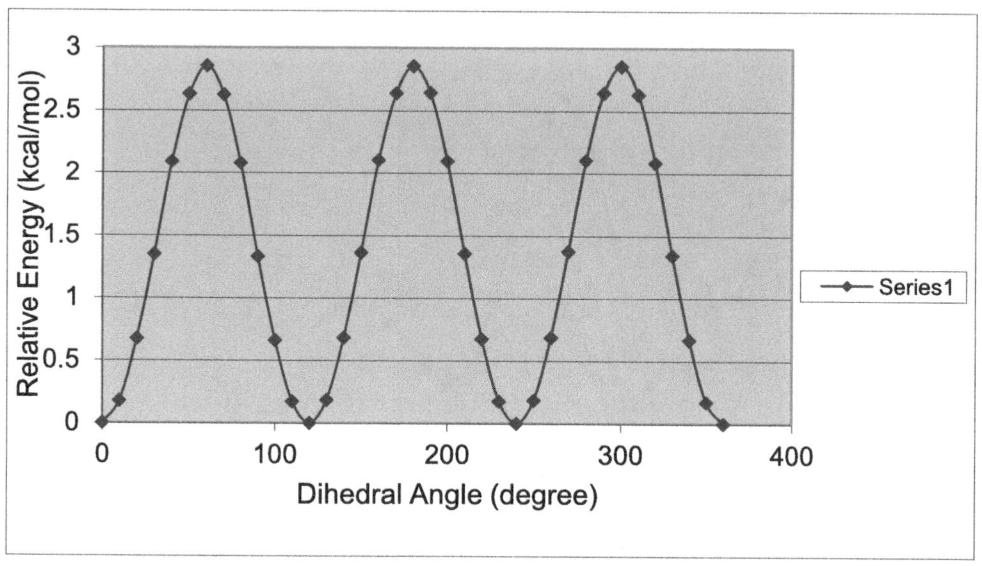

C-O Internal Rotor

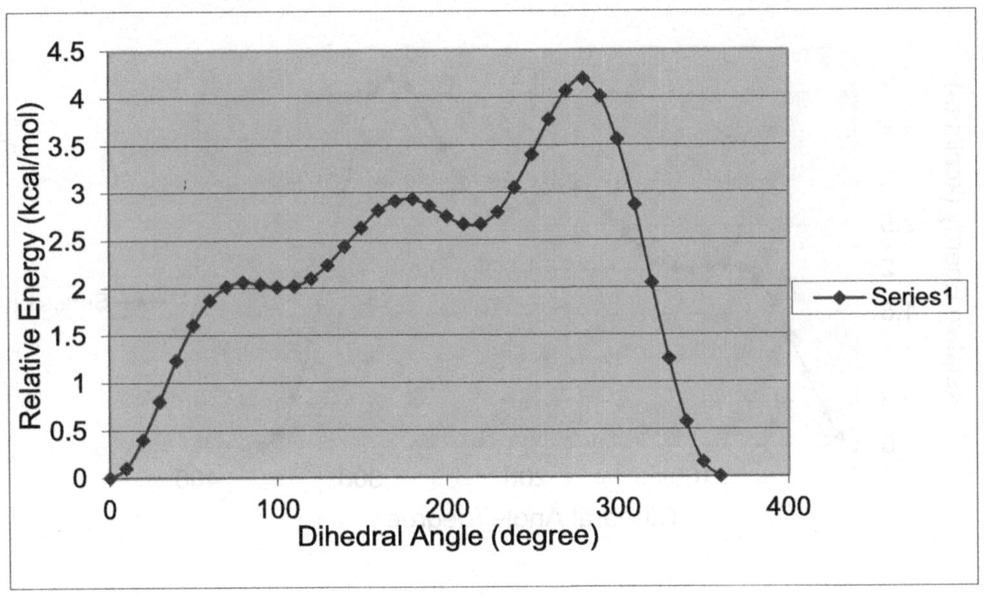

Figure 31. Potential energy profile of C-C and C-O internal rotors for CH2FC•FOH. The solid lines indicate Fourier series expansion

C-C Internal Rotor

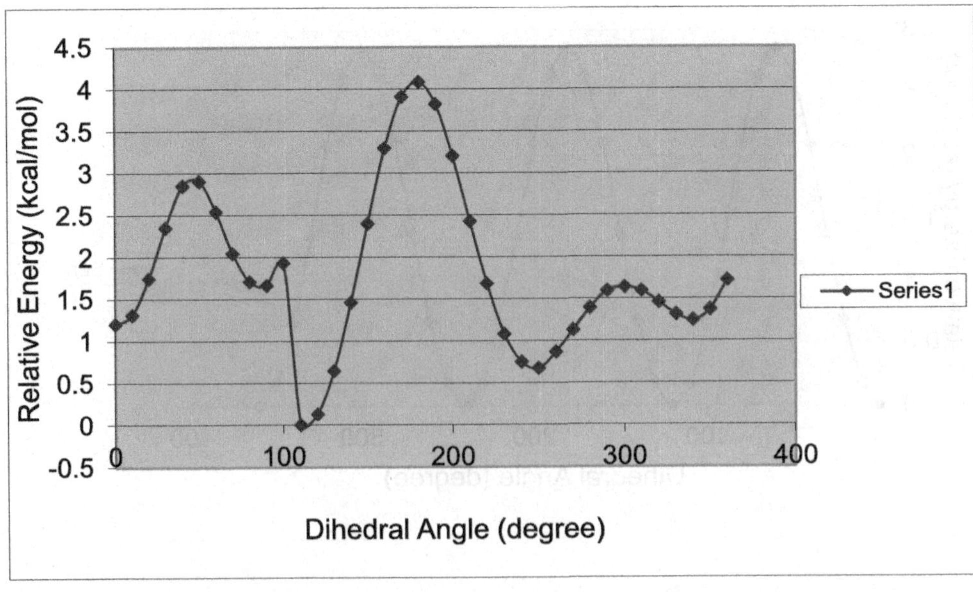

C-O Internal Rotor

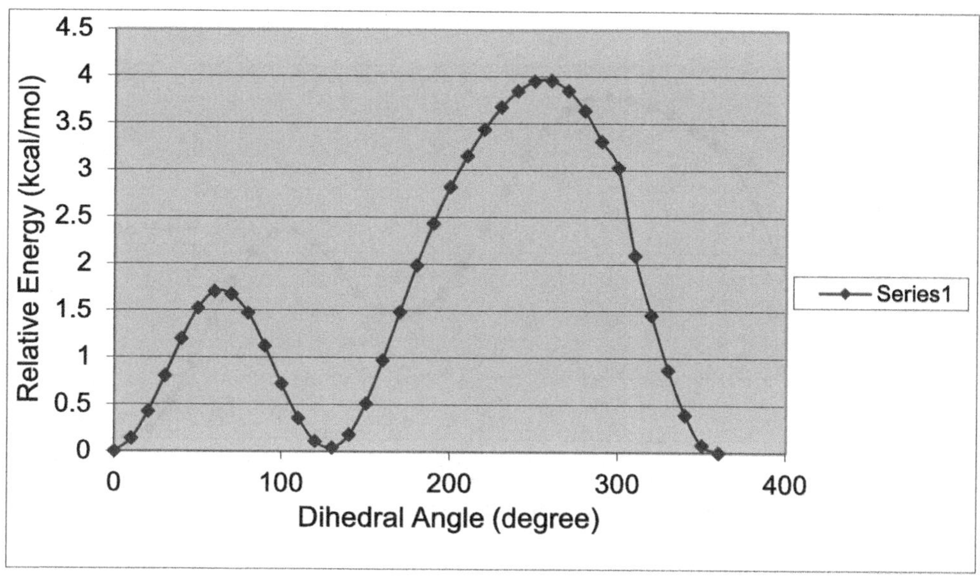

Figure 32. Potential energy profile of C-C and C-O internal rotors for C•HFCFHOH. The solid lines indicate Fourier series expansion

C-C Internal Rotor

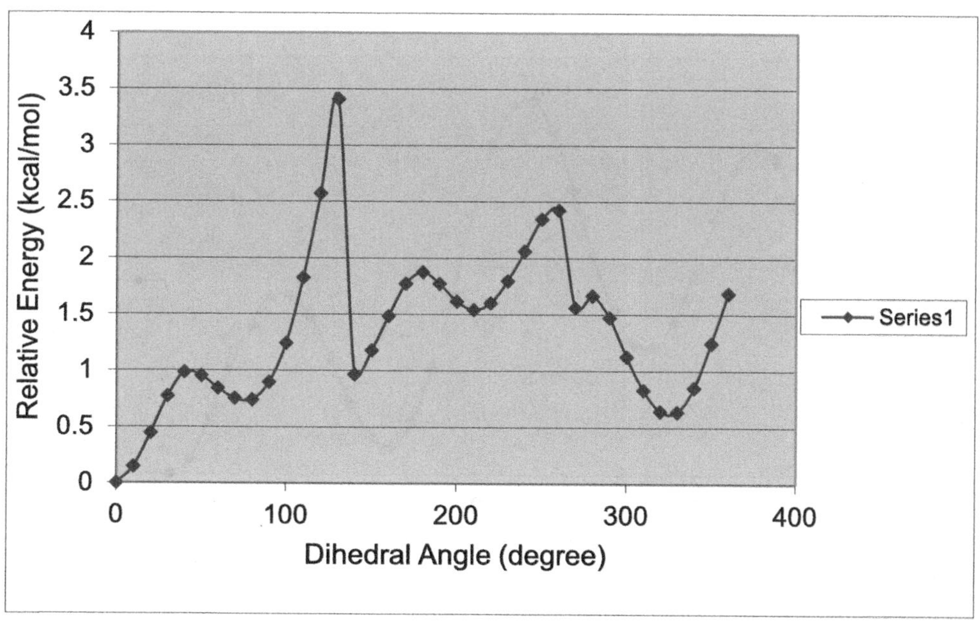

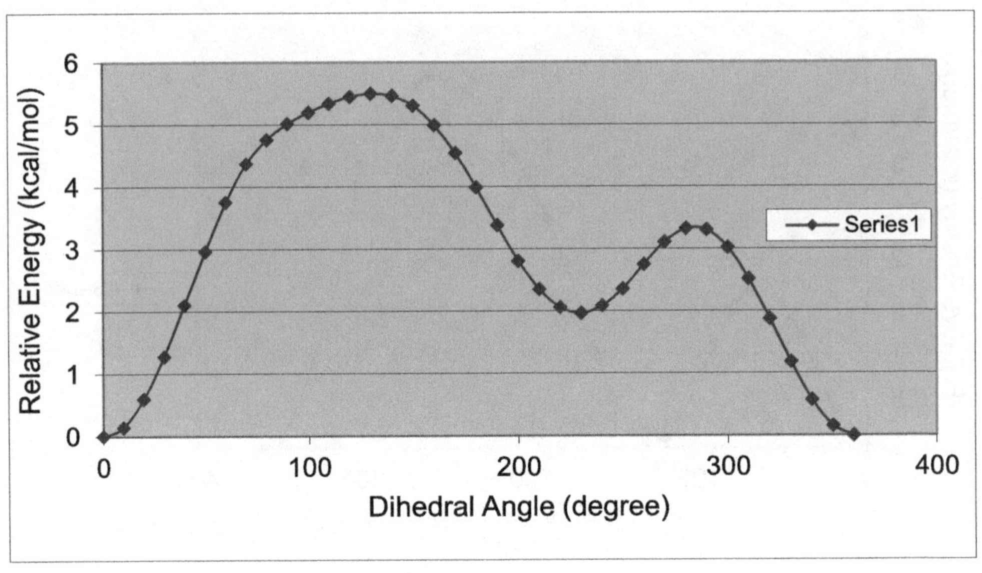

Figure 33. Potential energy profile of C-C and C-O internal rotors for CH2FCHFO•. The solid lines indicate Fourier series expansion

C-C Internal Rotor

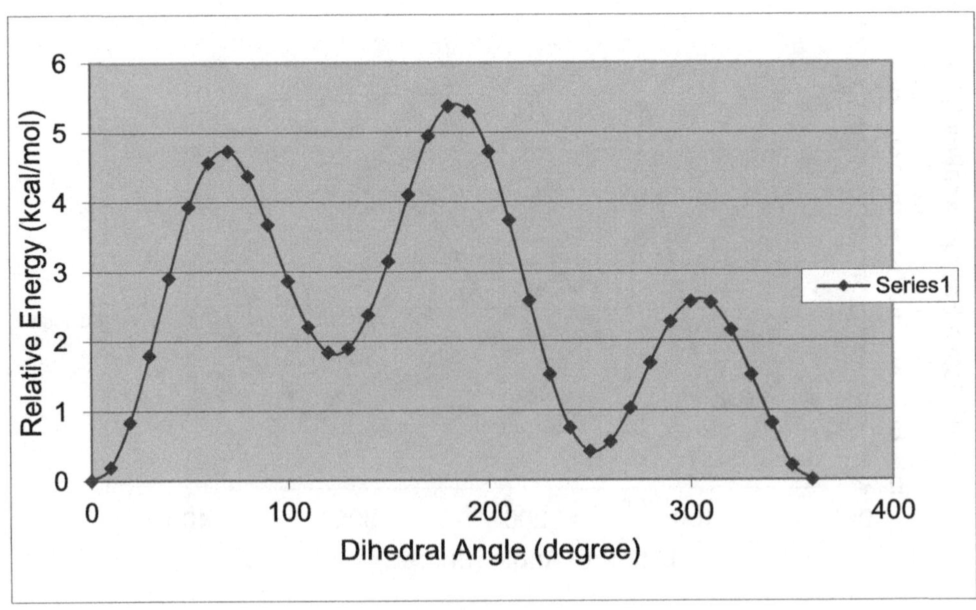

C-C Internal Rotor

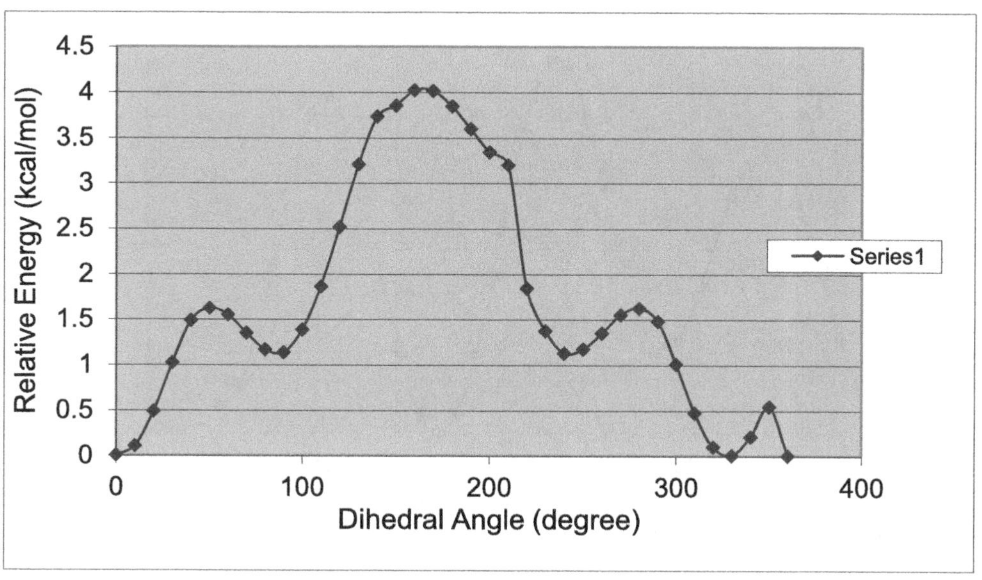

C-O Internal Rotor

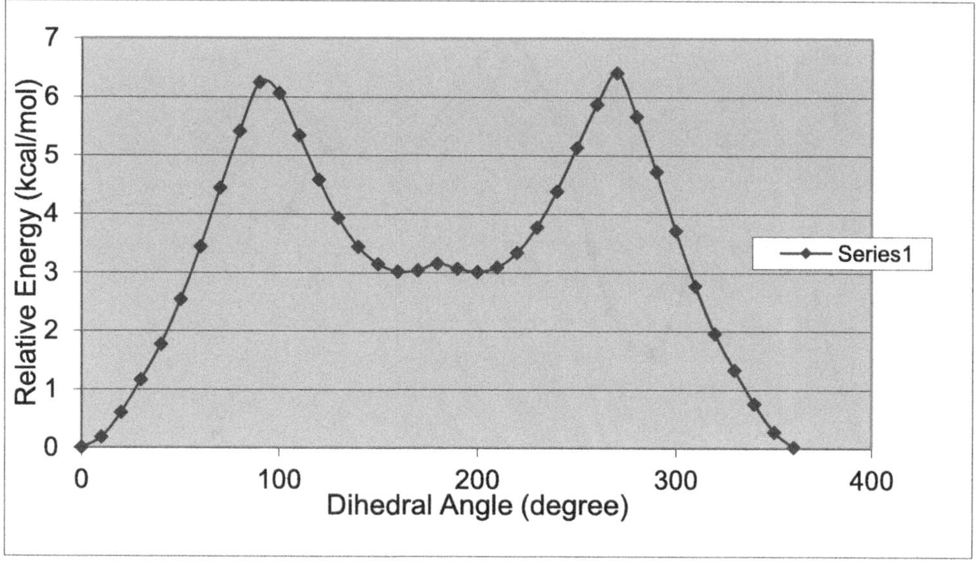

Figure 35. Potential energy profile of C-C and C-O internal rotors for C•F2CH2OH. The solid lines indicate Fourier series expansion.

C-C Internal Rotor

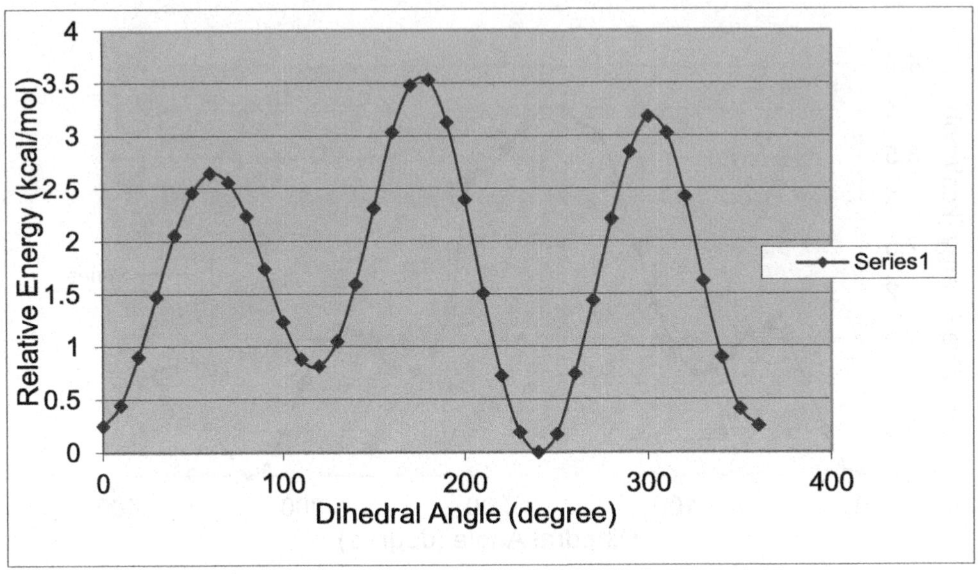

C-O Internal Rotor

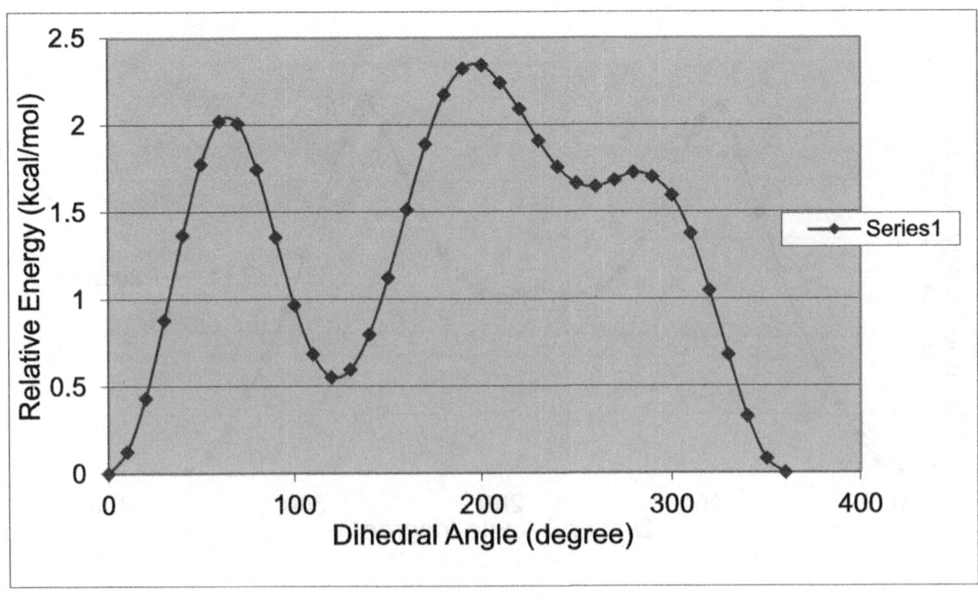

Figure 36. Potential energy profile of C-C and C-O internal rotors for CHF2CH2O•. The solid lines indicate Fourier series expansion.

C-C Internal Rotor

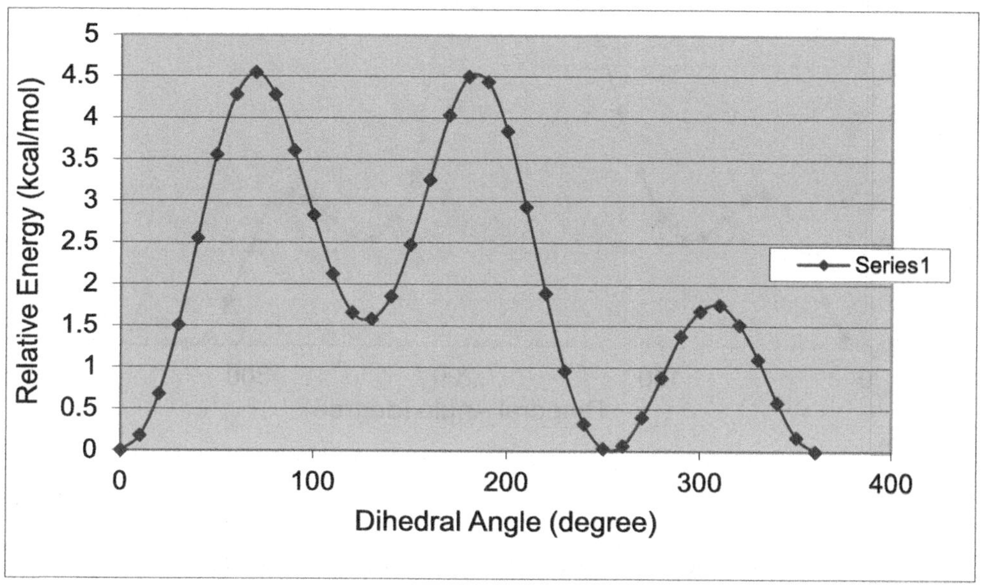

Figure 37. Potential energy profile of C-C and C-O internal rotors for CH2•CF2OH. The solid lines indicate Fourier series expansion.

C-C Internal Rotor

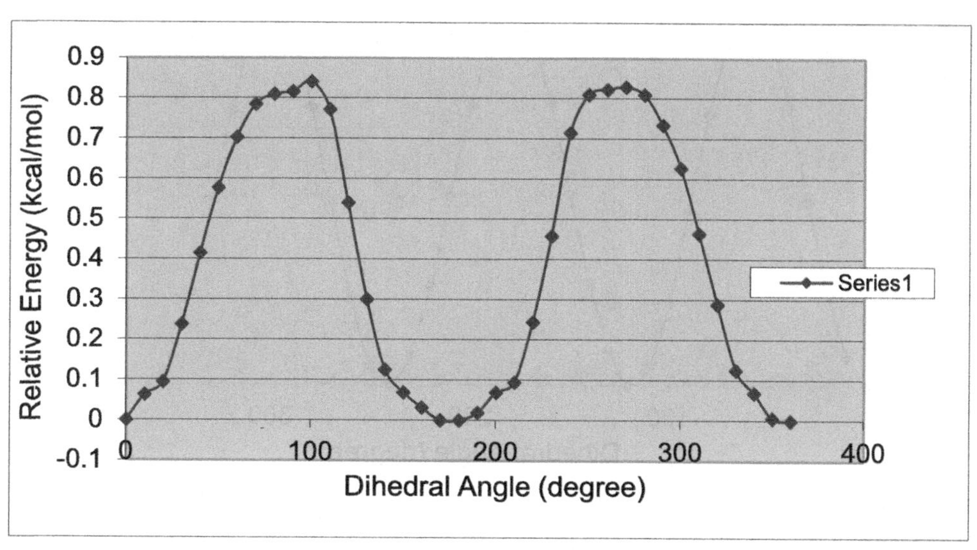

C-O Internal Rotor

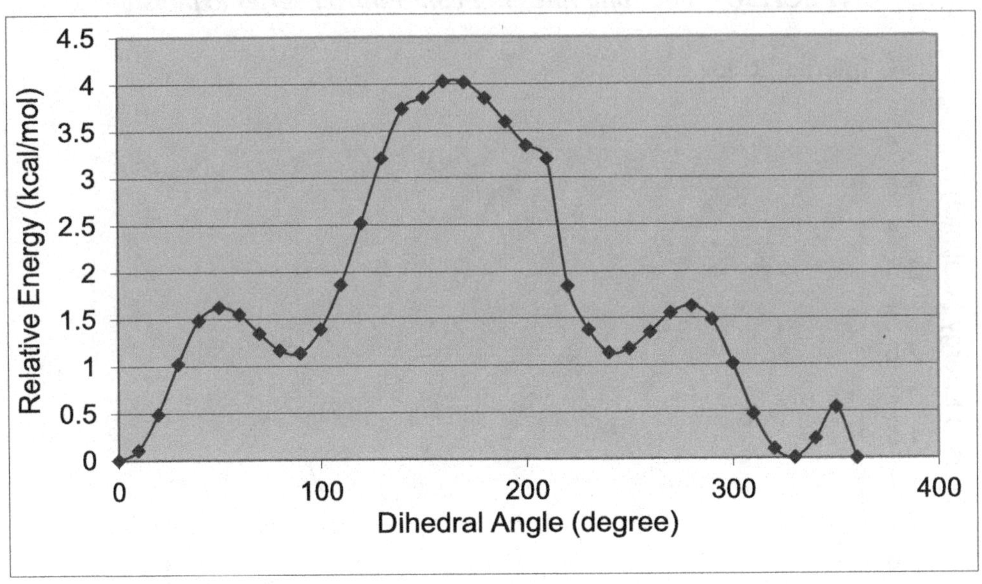

Figure 38. Potential energy profile of C-C and C-O internal rotors for CH3CF2O•. The solid lines indicate Fourier series expansion.

C-C Internal Rotor

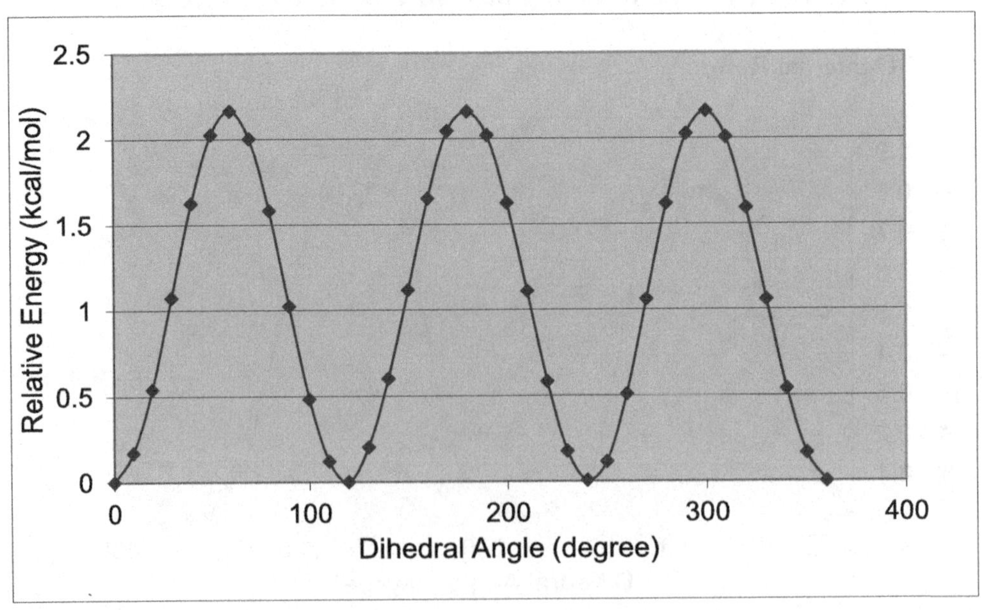

Table 4. The method of standard deviation and example calculation

The standard deviation is calculated using the following formula [36] :

$$\sigma = \sqrt{\frac{1}{N}\sum_{i=1}^{N}(x_i - \mu)^2}$$

Where, X_i is the mean; the average of the numbers, μ is the actual numbers to be calculated the standard deviation of, and

$$\frac{1}{N}\sum_{i=1}^{N}(x_i - \mu)^2$$ is the variance.

The standard deviation for standard enthalpy of formation for 1-fluroethanol is calculated to be ± 0.1. The calculation of standard deviation for this molecule is as follows:

N=2, μ_1= -101.60, μ_2= -101.86

Step 1, X_i= (-101.60-101.86)/2 = -101.73

Step 2, $(X_i - \mu_1) = (-101.73 + 101.60)^2 = 0.0169$

$(X_i - \mu_2) = (-101.73 + 101.86)^2 = 0.0169$

Step 3, ½ (0.0169 + 0.0169) = 0.0169

Step 4, square root of 0.0169 = 0.13= ± 0.1

Chapter 8
Thermochemistry and Reaction Kinetics of the Stabilized and the Chemically Activated Secondary Ethyl Radical of Methyl Ethyl Sulfide with Oxygen

1. Overview

This chapter presents the use of the quantum Rice-Ramsperger-Kassel (QRRK) theory method to analyze the reaction between the activated CH_3S $CH{\cdot}CH_3$ and molecular oxygen to account for further reaction and collisional and deactivation. Hydroxyl radicals initiate the oxidation of Methyl ethyl sulfide ($CH_3SCH_2CH_3$) and MES (methylthioethane) under combustion conditions. Stable and unstable products of oxidation of activated secondary radical of methyl ethyl sulfide is presented and is based upon the calculated kinetic and thermodynamic parameters.

The chapter also presents the thermochemical properties of reactants, products, and transition states using the CBS-QB3 and G3MP2B3 composite and M062X/6-311+G(2d, p) DFT. The calculated kinetic parameters using the thermochemical properties of products, reactants, and transition states under the CBS-QB3 method of calculation are also presented.

2. Introduction

Biofuels contain lots of sulfur-containing substances. In combustion science, the kinetics for the decomposition of sulfur compounds under different pressure and temperature play an important role. The reaction between O_2 and $CH_3S(OH)CH_3$ was studied by Gross et al. using DFT methods and *ab initio* theory. Information on the partial oxidation of sulfur hydrocarbons with three or more carbons and the kinetics of combustion is still in need.

The reaction mechanisms and kinetics for the destruction of different types of sulfur compounds under varied temperatures and pressure play an important role in combustion science. A better understanding of the combustion and partial oxidation of sulfur compounds is important for the reduction and prevention of sulfuric oxides that cause destruction to the environment, buildings, and human health. Gargurevich investigated the combustion mechanisms and kinetics of H_2S under Claus Furnace conditions based on the reviews of recently published reaction mechanisms and kinetics modeling. Gross et al. theoretically studied the reaction between $CH_3S(OH)CH_3$ and O_2 using *ab initio* theory and DFT methods. There is, however, still a need for information on the kinetics of combustion and for the partial oxidation of sulfur hydrocarbons with three or more carbons in the atmosphere. A significant amount of methyl ethyl sulfide (MES) and $CH_3SCH_2CH_3$ methylthioethane are emitted into the atmosphere due to biological activities in the sea. MES has a negative effect on the environment and atmosphere through its generation and contribution to the production of smog and sulfuric oxides. The oxidation of MES under combustion conditions and a higher temperature is of value for kinetic modeling.

Kinetics and Thermochemistry of C_2H_5 + O_2 and the chemically activated and stabilized $CH_3CH_2OO\bullet$ adduct was presented by Sheng and Bozzelli. Thermodynamic and kinetic analysis of the reaction of $CH_3SCH_2\bullet$ + O_2 was reported by Jin and Bozzelli. In 2014, Thermochemistry, kinetics, and reaction paths on the tert-isooctane radical reaction with O_2 were studied by Snitsiriwat and Bozzelli.

Structural and thermochemical properties of the stable molecules of the sulfur peroxides and alcohols of MES and all of its partial oxidation intermediates were previously studied by Song et. al.

In 1994 Fu and group studied reaction kinetics and thermochemistry for the oxidative polymerization reaction between ammonium peroxy-persulfate (APP) and Aniline (AN). The product polyaniline is a member of conducting polymer family and has some industrial applications. The use of literature bond energies to calculate the enthalpy of this reaction was limited due to some structural features of the conducting polyaniline, such as protonation and conjugation. Using literature bond energies to calculate the enthalpy of the reaction would be inaccurate, neglecting polymer backbone protonation and conjugation. It's known that the oxidative polymerization reaction of Aniline (AN) by a chemical oxidant is highly exothermic.

The study of thermochemistry as a function of the nature of acid, PH, initial concentration of aniline, and concentration ratio APS/AN was performed to propose a mechanism for this oxidative polymerization reaction. The enthalpy change for the reaction was found to be -105 kcal/mol. The type of product formed was found to be dependent on the concentration of acid and the ratio of concentrations APS/AN. Using thermograms for the polymerization, it's postulated that the formation of aniline dimer from the aniline monomers is the rate-determining step in neutral media. Thermograms for the polymerization in the presence of acid show that polymerization occurs in one step and occur slowly and in two steps in neutral media. UV-vis spectra show that the low molecular weight of aniline oligomers is formed in neutral media, and the emeraldine form of polyaniline is formed in acidic media.

This chapter presents the intramolecular isomerization (hydrogen transfers), beta scissions, and intramolecular addition reaction of the peroxy oxygen radical to the sulfur atom, resulting in peroxy oxygen–oxygen and the methyl carbon-sulfur bond cleavages and in oxygen–sulfur and oxygen–carbon π bond formation on the sulfur atom in the carbon components.

3. Experimental and Data

Thermochemical properties of the parent molecules and their radical fragments were calculated using several density functional theories and composite *ab initio* computation chemistry methods in the Gaussian 09 program suite. The different levels of calculation used in this study include the M06-2x/6-311+g(2d,p) density functional theory (DFT) method and the *ab initio* CBS-QB3 and G3MP2B3 composite methods. The composite CBS-QB3 calculation method has been used for the thermochemical properties in order to achieve improvements in the accuracy over the DFT methods. The CBS-QB3 method utilizes the B3LYP/6-311G(2d,d,p) level of theory to get the optimized lowest energy geometries and to calculate frequencies; it then uses CCSD(T), MP4(SDQ), and MP2 calculations to determine single point energies. The G3MP2B3 method is a modified version of the G3MP2 method that uses the geometries and vibrational energies obtained from B3LYP/6-31G(d) calculations. The energy of each transition state was calculated based on the energy of the corresponding reactant plus the calculated energy difference between the transition state and the reactant.

Molecular structures and vibrational frequencies were determined at the B3LYP/6-311G (2d,d,p) in the CBS-QB3 level of calculation, which is considered accurate for the calculation of electronic structure and energies of the first and second-row atoms. The Complete Basis Set-QB3 multi-level method was developed by the research group of G. Peterson and Kiselev *et al*. The CBS-QB3 multilevel method is based on the optimized geometries at the B3LYP/6-311G(2d,d,p) level.

The entropy and heat capacity of molecules, radicals, and transition states as a function of temperature were from the optimized structures, moments of inertia, vibration frequencies, symmetries, electronic multiplicity, and molecular mass. All of the optimized lowest energy structures of reactants and products were identified as having no imaginary frequency. Transition states were identified with one imaginary frequency related to the vibration that demonstrates the transitional motion between the reactant and the product. Contributions of translation, external rotation, and vibrations were determined with standard formulas from statistical mechanics. Contributions to the entropy and heat capacity of each molecule and radical from translation, vibrations, and external rotation were determined using the Statistical Mechanics for Heat Capacity and Entropy, C_p and S (SMCPS) program. The SMCPS program utilizes the rigid-harmonic oscillator approximation from the optimized structures obtained at the B3LYP/6-311G(2d,d,p) level. The number of optical isomers, when they exist, and the spin degeneracy of unpaired electrons were also incorporated into the calculation of S°_{298K}. The SMCPS program utilizes the rigid-harmonic oscillator approximation from the optimized structures obtained in the CBS-QB3 calculation. The scaling factor for zero-point vibrational energy is 0.99.

The internal rotor potential energy diagrams and the lowest energy-optimized conformer for each target, parent peroxide molecule, and radical were determined using the B3LYP/6-31+G(d,p) level DFT calculations. Internal rotor analysis was performed on each C—S, C—C, C—O, or O—O single bond rotor to determine the lowest energy structures. All rotors were re-scanned once a lower energy conformer was found compared to the initial energy conformer until the lowest energy geometry was identified.

Contributions from all of the hindered internal rotors to S°_{298} and $C_p(T)$ were determined using the Rotator program. The Rotator program calculates the thermodynamic functions from hindered rotations with arbitrary potentials based on the method developed by Krasnoperov, Lay, and Shokhirev. This technique employs expansion of the hindrance potential in the Fourier series, calculation of the Hamiltonian matrix on the basis of the wave functions of free internal rotors, and subsequent calculation of energy levels by direct diagonalization of the Hamiltonian matrix. Internal rotor torsion frequencies were omitted from the SMCPS frequency sets, and internal rotor contributions were added separately. The contributions from the identified torsion frequencies were omitted and replaced with the hindered rotor contributions. The moments of inertia for each of the two components of the rotor were calculated. The summation of the SMCPS and Rotator contributions gives the total entropy and heat capacity for each species.

The kinetics of O_2 addition to the radical sites of methyl ethyl sulfide and their reverse reactions were determined based on the Variational Transition State Theory (VTST). The M06-2x/6-311+g(2d,p) level C—OO• bond length scans were performed with the calculation of optimized structure, energy, and vibration frequencies at each 0.05 Å step. A complete set of thermochemical properties are calculated for the different transition state structures at each step. Rate constants are subsequently calculated from the reactant to each structure at temperatures of 300 to 2000 K. The minimum rate constant is taken across the temperature–bond length data set for each temperature. The minimum set of rate constants over the temperature range is fit to the modified Arrhenius rate constant form.

High-pressure limit reaction rate constants (k_∞, s^{-1} or $cm^3\ mol^{-1}\ s^{-1}$) were determined based on the canonical transition state theory. The values of k_∞

were fit to the three-parameter form of the Arrhenius equation to yield the parameters of A, n, and E_a over the temperature range of 200-3000K:

$$k_\infty = \Gamma(T) * A * (T^\wedge n) * \exp(-E_a/RT).$$

Where activation energies of reactions (E_a) are calculated as $E_a = [\Delta H_f^o{}_{298}\,(\text{TS}) - \Delta H_f^o{}_{298}\,(\text{Reactant})]$.

The quantum mechanical tunneling correction (Γ, dimensionless) was calculated using Eckart's method;

The entropy differences between the reactants and the transition states ($\Delta S^{\neq}$) over the temperature range of 200-3000 K were used to determine the pre-exponential factor (A, s^{-1}, or cm^3 mol^{-1} s^{-1}). The values of A are expressed by:

$$A\,(\text{T}) = (\kappa_b T/h_p)\exp(\Delta S^{\neq}/R)$$

Where h_p is Planck's constant and κ_b is Boltzmann's constant, R is the gas constant, and R = 1.987cal $mol^{-1}K^{-1}$.

The high-pressure limit parameters are calculated based on the thermochemical properties of reactant, product, and transition state geometries optimized under the CBS-QB3 level. Temperature and pressure-dependent rate constants are calculated using the multichannel, multifrequency Quantum Rice-Ramsperger-Kassel (QRRK) analysis for $k(E)$ based on a master equation analysis for falloff and stabilization as implemented in the CHEMASTER code. Reduced sets of three vibrational frequencies and their degeneracy plus energy levels of one external rotor are used to yield the ratio of the density of states to partition the coefficient, $\rho(E)/Q$, for each adduct (isomer in the chemical activation or dissociation reaction system). The Master Equation analysis used an exponential-down model for the energy transfer function with ($\Delta E_{down}{}^o$) 900 cal mol^{-1} for N_2 as the third body. Rate constants, k(E), were evaluated by using incremental energy bins of 1.0 kcal mol^{-1} up to 75 kcal mol^{-1} above the highest barrier. Lennard-Jones parameters, $\sigma(\text{Å})$, and ε/k (K) were obtained from tabulations and from an estimation method based on molar volumes and compressibility. High-pressure-limit elementary-rate parameters were used as input data for the QRRK calculations.

4. Results and Analysis

4.1 Reaction Pathways and Geometries of Optimized Transition States

According to the recent studies by Mansilla *et al.* and Cao *et al.*, the partial oxidation of MES in th from each MES molecule by one hydroxyl radical:

$$CH_3SCH_2CH_3 + HO\cdot \rightarrow CH_2\cdot SCH_2CH_3 + H_2O$$

$$CH_3SCH_2CH_3 + HO\cdot \rightarrow CH_3SCH\cdot CH_3 + H_2O$$

$$CH_3SCH_2CH_3 + HO\cdot \rightarrow CH_3SCH_2CH_2\cdot + H_2O$$

The radicals formed after the loss of one hydrogen atom from different carbon sites will subsequently react with O_2 to form active sulfuric alkyl peroxy radicals:

$$CH_2\cdot SCH_2CH_3 + O_2 \rightarrow CH_2(OO\cdot)SCH_2CH_3$$

$$CH_3SCH\cdot CH_3 + O_2 \rightarrow CH_3SCH(OO\cdot)CH_3$$

$$CH_3SCH_2CH_2\cdot + O_2 \rightarrow CH_3SCH_2CH_2OO\cdot$$

This chapter focuses on the thermochemistry, reaction paths, and kinetics of the $CH_3SCH\cdot CH_3$ radical reaction with O_2 and subsequent reactions.

The elementary reaction steps investigated in this paper belong to the following categories:

- Forward association and reverse dissociation reactions of the $CH_3SCH\cdot CH_3$ radical with O_2 are investigated by VTST;
- Intramolecular hydrogen transfer from carbon the three carbon sites to the peroxy oxygen radical (TS211, TS231); plus the H-transfer

from the ipso carbon to the peroxy oxygen, which further reacts with no barrier to form an aldehyde plus hydroxyl radical (TS221);

- Molecular dissociation caused by C—S bond cleavage (TS213, TS234);
- Reactions to cyclic ether ring formation plus OH radical (TS212, TS232);
- Intramolecular addition of peroxy oxygen radical to the sulfur atom with simultaneous cleavage of the weak O—O and C—S bonds, resulting in the information of two products with C=O and S=O double bonds (TS251);
- HO_2 elimination to generate CSC=C (TS233, TS241)

Figure 1: Subsequent reaction steps of $CH_2SCH(OO\bullet)CH_3$ radical along with the numbering of the corresponding transition states.

$CH_3SCH(OO\bullet)CH_3$ ←TS211→$H_2C\bullet SCH(OOH)CH3$ --TS212→ $H_2\overset{O}{\overset{\triangle}{CSC\text{-}C}}H_3$+ OH
    --TS213→ $CH_2S + OH + CH_3CH{=}O$

$CH_3SCH(OO\bullet)CH_3$ --TS221→ $CH_3SC\bullet(OOH)CH_3$ → $CH_3SC({=}O)CH_3 + OH$

$CH_3SCH(OO\bullet)CH_3$ ←TS231→ $CH_3SCH(OOH)CH_2\bullet$ --TS232→$H_2\overset{O}{\overset{\triangle}{CSC\text{--}C}}H_2$+ OH
    --TS233→$CH_3SCH{=}CH_2$+ HO2
    --TS234→ $CHS\bullet + CH_2{=}COOH$

$CH_3SCH(OO\bullet)CH_3$ --TS241→ $CH_3SCH{=}CH_2 + HO_2$

$CH_3SCH(OO\bullet)CH_3$ --TS251→ $CH_3CH{=}O + CHS^{\cdot}{=}O$

$CH_3SCH(OO\bullet)CH_3$ ←VTST→$CH_3SCH\bullet CH_3 + O_2$

The optimized lowest energy structures and thermochemical properties of MES, along with its radicals and the peroxide molecules and radicals, have already been previously studied.

Transition states are characterized as having only one negative eigenvalue of Hessian (force constant) matrices. The absence of imaginary frequency verifies that the structures are true minima at the respective levels of theory. The optimized lowest energy transition state conformers from the CBS-QB3 level are drawn in Figure 2.

Figure 2: Geometry of lowest energy conformer of transition states at 298K calculated under CBS-QB3 level.

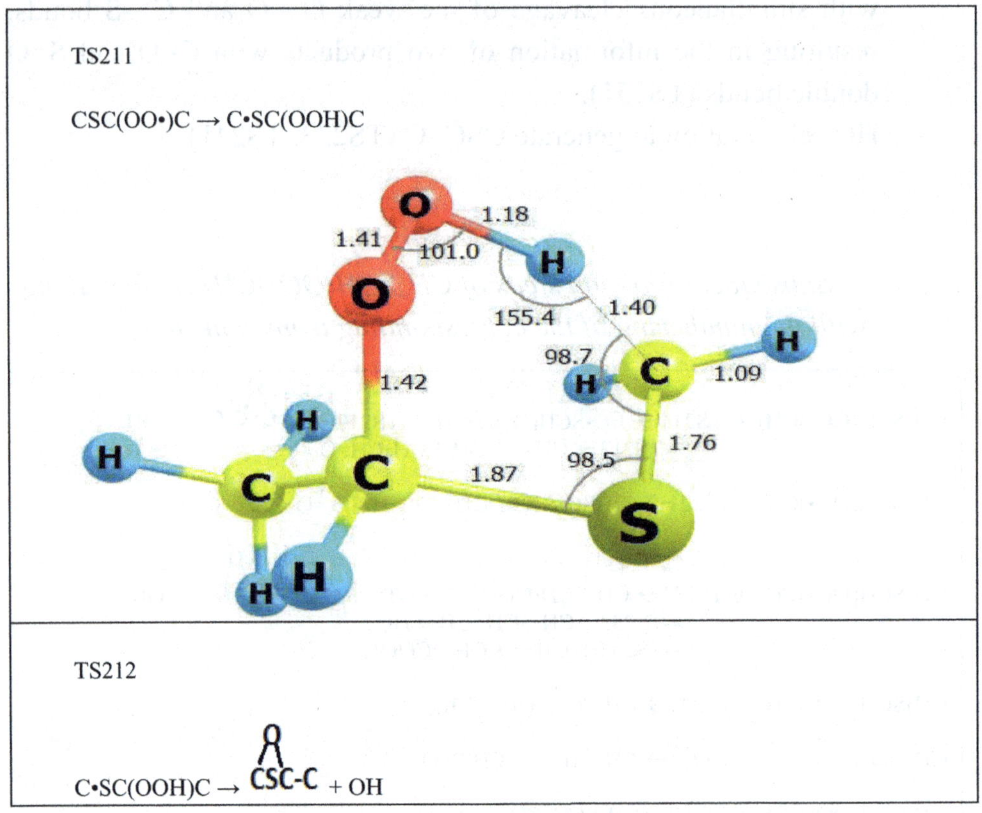

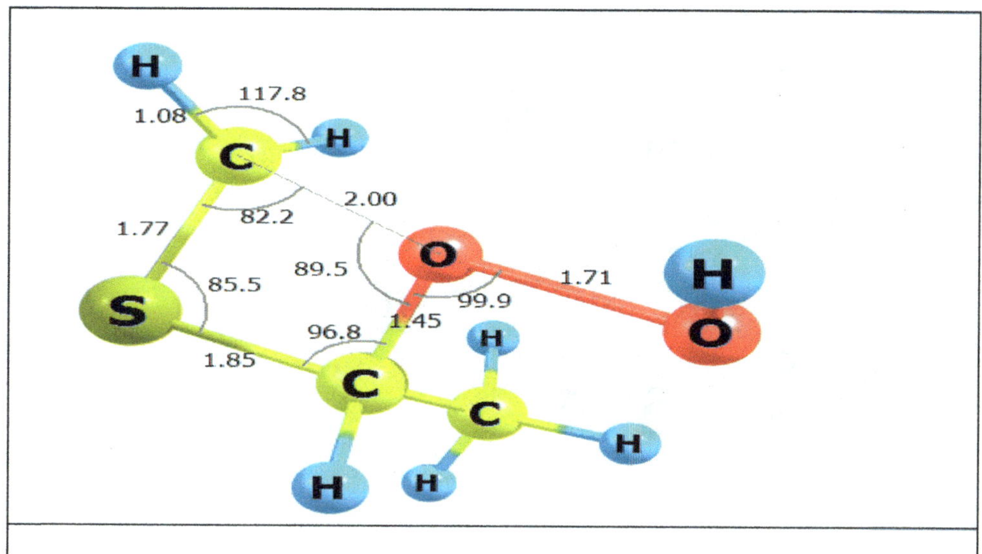

TS213

C•SC(OOH)C → CH$_2$=S + CC=O + OH

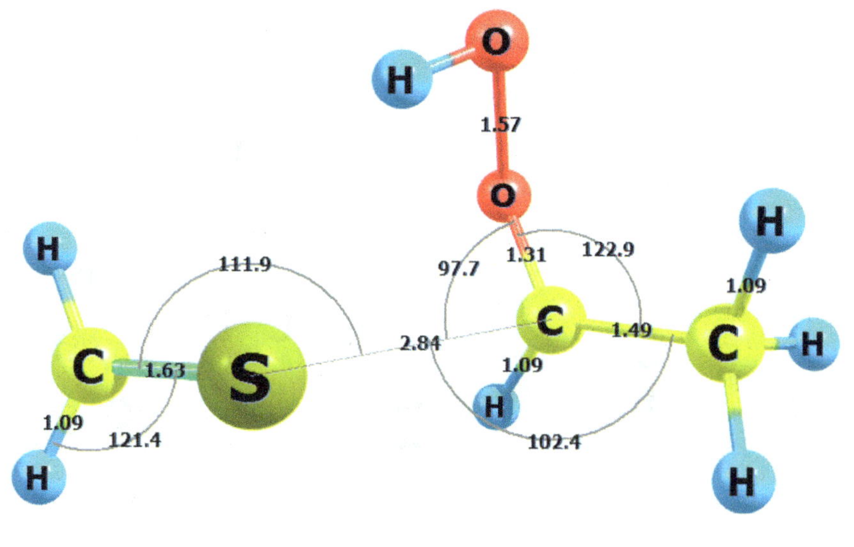

TS221

CSC(OO•)C → CSC(=O)C + OH

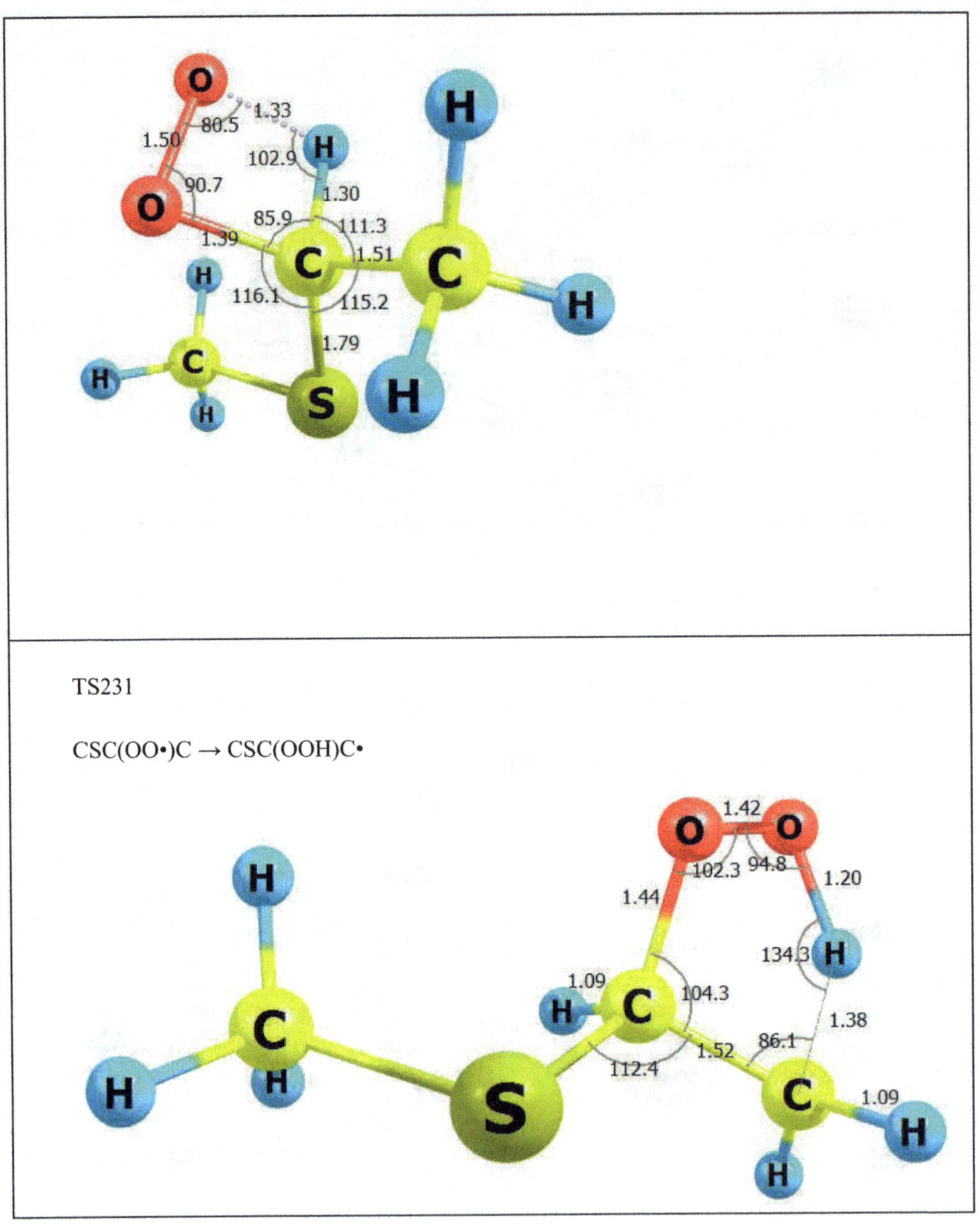

TS231

CSC(OO•)C → CSC(OOH)C•

CSC(OOH)C• → CSC--C + OH

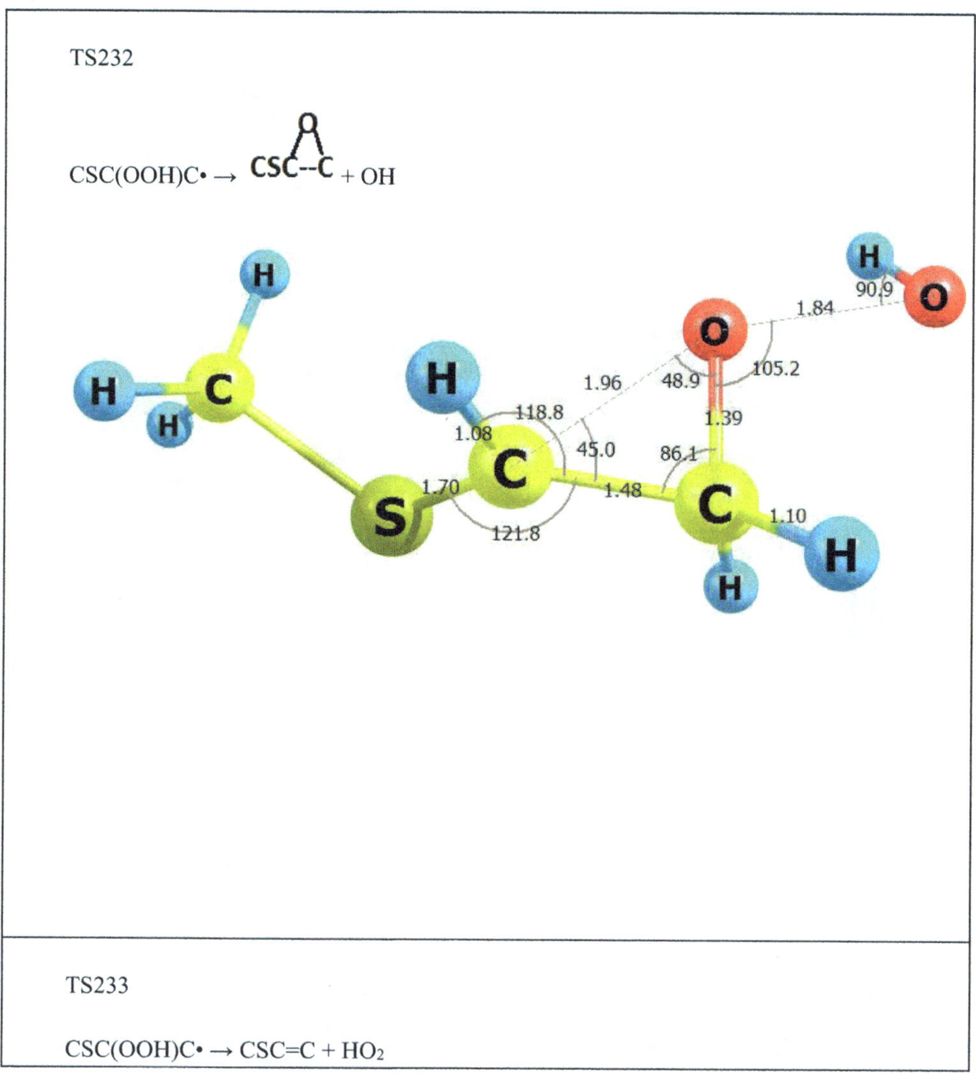

TS233

CSC(OOH)C• → CSC=C + HO_2

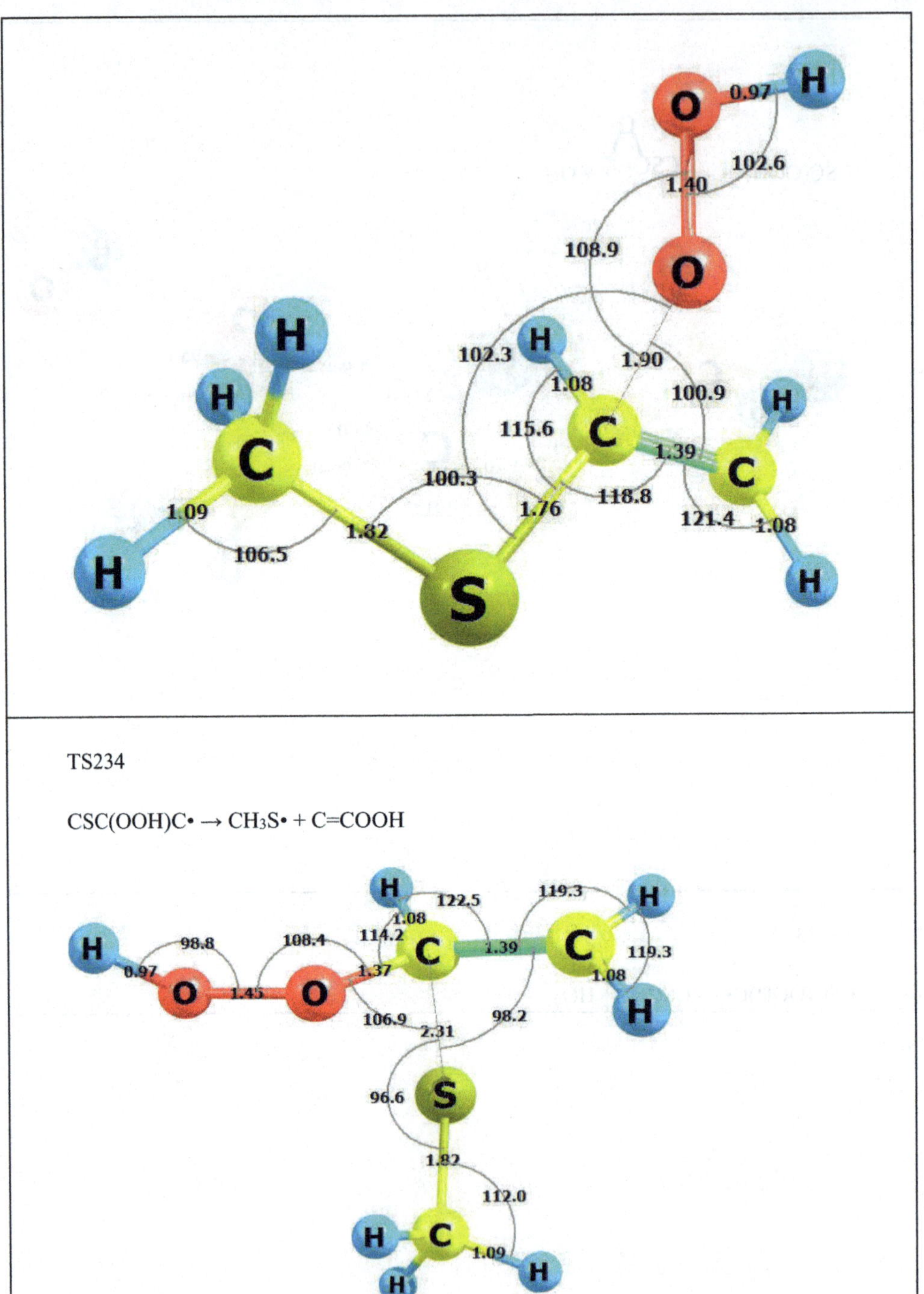

TS234

$CSC(OOH)C\bullet \rightarrow CH_3S\bullet + C=COOH$

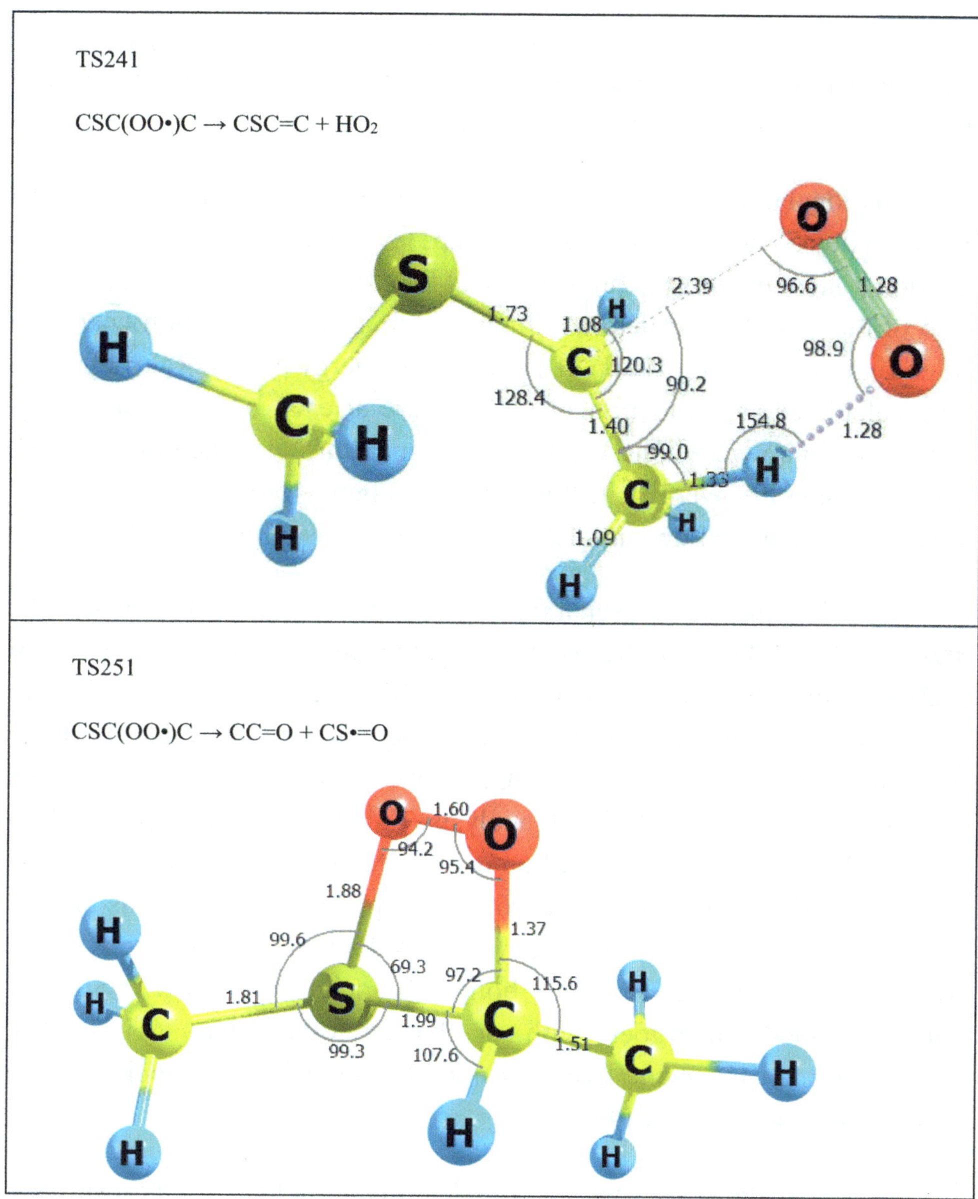

4.2 Variational Transition State Theory (VTST) Analysis

A Zero-point corrected potential energy surface for the RC—OO• bond is illustrated in Figure 3. Potential energy scans were performed along the R—OO• → R• + O2 dissociation path in the MES peroxide bond cleavage reaction at 0.05 Å intervals. Frequency and thermodynamic calculations were performed at each of the points along the potential energy surface. Thermochemical properties and rate constants as a function of temperature were calculated at each point along the potential energy surface.

Variational transition kinetics were calculated based on thermochemical data obtained at the M062X/6-311+G(2d,p) level. Each point on the M062X reaction potential energy scan was scaled by a factor that accounted for the ratio of the CBS-QB3 reaction enthalpy to the M062X reaction energy in order to improve the precision of dissociation energy points. Rate constants were calculated as a function of temperature at each bond distance. The overall minimum rate constant at each temperature was then determined. Rate constants (T) were fit to the three-parameter form of the Arrhenius equation to yield the rate parameters A, n, and E_a listed in Table 1.

The VTST plot of O_2 separation from peroxide radical indicates the BDE of •OO—$CH_2SCH_2CH_3$ is about 30.0 kcal/mol, close to the value of 30.2 kcal/mol that is indicated by the PE diagram in Figure 4. For comparison to hydrocarbons, Sheng et al. reported values for CH_3CH_2—OO• BDE of 35.32 kcal/mol by CBS-QB3 and 35.5 kcal/mol by and B3LYP/6-31G(d, p) level VTST analysis. Snitsiriwat et al. reported R—COO• BDE values in tertiary carbon of -isooctane hydroperoxide based on CBS-QB3 calculation and B3LYP/6-31G(d, p) level VTST scans at 34.7 kcal/mol and 33.5 kcal/mol, respectively. Similar to the dissociations of Jin, Sheng, and Snitsiriwat's C—OO• bonds, dissociation of the CH_2SCH(—OO•)CH_3 bond undergoes with no barrier other than ΔHrxn. For Snitsiriwat's tert-isooctane hydroperoxide C—OO• bond dissociation, the TS was 3.0 Å at T=300 K.[48]

The normalized energy versus bond length plot in Figure 3 indicates that a transition state occurs between the •OO—$CH_2SCH_2CH_3$ bond length of 2.0 ~ 2.3 Å. Based on the analysis of the rate constants of both the association and dissociation reactions, the M062X method was able to converge on an optimized TS geometry with the C—OO• bond length of 2.15 Å for CH_2SCH(--OO•)CH_3. It appears that compared to saturated hydrocarbons, with only carbon and hydrogen, the presence of sulfur reduces the chemical activation energy and the transition state C—OO•

bond length for O_2 association to carbon atoms adjacent to the sulfur moiety by about 7 kcal/mol and 0.8 Å, respectively.

Figure 3 The CH₂SCH(—OO•)CH₃ bond length scan for VTST calculation.

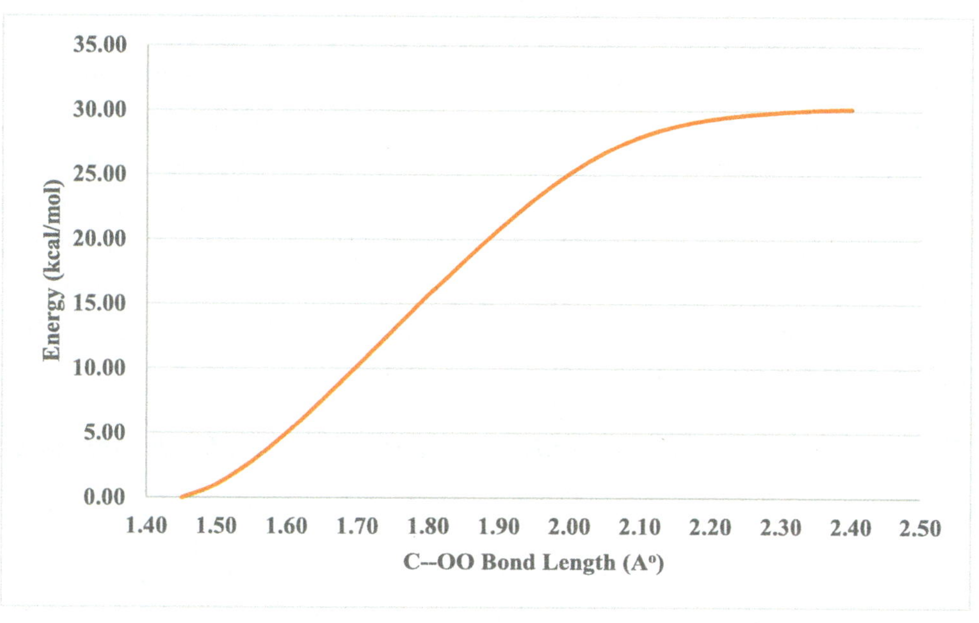

Table 1 Elementary rate parameters from VTST analysis for CH₂SCH(OO•)CH₃.

Association			Dissociation		
A^a	n	E_a (kcal)	A^b	n	E_a (kcal)
2.86E+3	2.24	-1.61	8.80E+12	0.33	29.41

[a]The unit of pre-exponential factor for the association reaction is cm^3 mol^{-1} s^{-1}.

[b]The unit of the pre-exponential factor for the dissociation reaction is s^{-1}.

4.3 Thermochemical and Kinetic Analysis

The energy barriers (E_a) and reaction enthalpy (ΔH_{rxn}) for the reactions of $CH_2SCH(OO\bullet)CH_3$ under the M06-2X/6-311+G(2d,p), G3MP2B3 and CBS-QB3 calculation levels are presented in Table 2. The energy of each transition state structure is calculated from the corresponding reactant plus the energy difference between the transition state and the reactant. The CBS-QB3 calculation shows good agreement with the M062X/6-311+G(2d,p) calculation in the intramolecular hydrogen shift and β-scission reactions. There is, however, a 4 ~ 7 kcal/mol difference in the E_a and ΔH_{rxn} values of the reactions involving ring formation and for $O\bullet$ attachment to S., The two composite methods CBS-QB3 and G3MP2B3 agree reasonably well for ΔH for each of the reactions and TST's.

Table 2 E_a and ΔH_{rxn} values under different calculation levels (kcal/mol)

Reactions	CBS-QB3		M062x/6-311+G(2d,p)		G3MP2B3	
	E_a	ΔH_{rxn}	E_a	ΔH_{rxn}	E_a	ΔH_{rxn}
$CSC(OO\bullet)C \rightarrow C\bullet SC(OOH)C$	18.3	8.3	20.7	12.1	20.4	10.3
$C\bullet SC(OOH)C \rightarrow$ CSC-C (ring, =O) $+ OH$	26.0	-14.3	30.9	-16.1	27.1	-16.1
$C\bullet SC(OOH)C \rightarrow CH_2{=}S + CC{=}O + OH$	22.1	-8.7	22.5	-7.6	24.9	-12.1
$CSC(OO\bullet)C \rightarrow CSC({=}O)C + OH$	36.3	-36.3	39.5	-35.3	38.7	-38.3
$CSC(OO\bullet)C \rightarrow CSC(OOH)C\bullet$	34.6	16.6	37.7	17.0	-----	16.6
$CSC(OOH)C\bullet \rightarrow$ CSC--C (ring, =O) $+ OH$	7.8	-14.9	14.4	-15.8	10.9	-16.2
$CSC(OOH)C\bullet \rightarrow CSC{=}C + HO_2$	15.3	6.2	18.0	6.3	16.3	4.3
$CSC(OOH)C\bullet \rightarrow CH_3S\bullet + C{=}COOH$	9.0	8.5	9.8	8.8	10.6	7.2
$CSC(OO\bullet)C \rightarrow CSC{=}C + HO_2$	28.5	22.8	30.1	23.3	27.7	20.9
$CSC(OO\bullet)C \rightarrow CC{=}O + CS\bullet{=}O$	31.1	-53.3	38.9	-46.8	32.2	-53.3

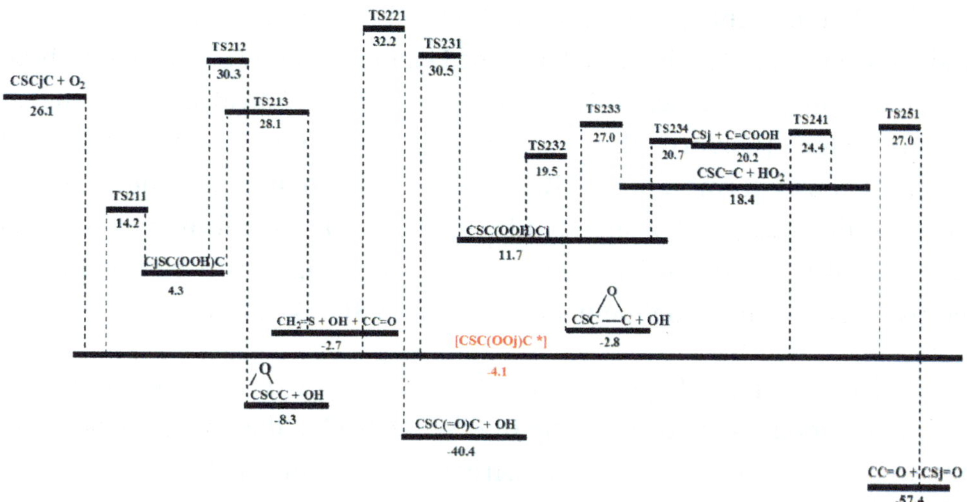

The potential energy diagram in Figure 4 is based on the CBS-QB3 level potential energy calculations for each species. The enthalpies of formation for the peroxides, radicals, aldehydes, and ketones of MES are from our previous work, while those of O_2, OH, and HO_2 are from Ruscic *et al*. The total enthalpies of products for each reaction step in Figure 4 are calculated based on the difference between the CBS-QB3 total energies of the reactants and the products.

TS211 in Figure 4 illustrates the intramolecular transfer of the H atom from the methyl carbon of CSC(OO•)C to the peroxy radical site through a six-member ring TST structure. The reaction results in the formation of C•SC(OOH)C and has a barrier of 18.3 kcal mol⁻¹.

- The methyl radical on this newly formed C•SC(OOH)C intermediate can undergo a substitution reaction into the carbon-bonded oxygen of the hydroperoxide to form a cyclic four-member ring 'thiol–ether' with the elimination of an OH radical. The reaction barrier is 26.0 kcal mol⁻¹.

- The C•SC(OOH)C radical can also undergo C•S--C(OOH)C bond beta scission to form CH_2=S and CC•OOH over a barrier of 23.8 kcal/mol. The CC•OOH radical is unstable and instantaneously undergoes β-scission forming a CC=O π bond and CC=O—OH scission to the lower energy products CC=O and OH.

TS221 represents the formation of the CSC(=O)C ketone plus hydroxyl radical via intramolecular transfer of the H atom from the secondary carbon, i.e., the carbon of the peroxide group to the peroxy radical site, forming an unstable the [CSC•(OOH)C]*. The CSC•(OOH)C is unstable and instantaneously forms a new CSC=O π bond of about 80 kcal mol-1; while cleaving the weak (~ 45 kcal mol-1 [CSC=O—OH] bond forming CSC(=O) ketone and OH radical. This ketone formation path has a barrier for the peroxy oxygen attack on the ipso hydrogen of 36.3 kcal/mol.

TS231 involves the intramolecular transfer of an H atom from the primary carbon on the ethyl group of CSC(OO•)C to the peroxy radical site through a five-member ring TST structure, resulting in the formation of CSC(OOH)C•.

- The TS231 has a 34.6 kcal mol^{-1} barrier. The higher barrier is a result of strain in the *5-member ring* and the stronger C—H bond of this primary carbon site on the ethyl group.
- Following this intramolecular hydrogen transfer, the newly formed methyl radical [CSC(OOH)C•] can undergo substitution into the peroxide oxygen, TS232. This has a low barrier of 7.8 kcal/mol for the formation of a 4-member epoxy ring, CSy(COC) plus OH radical.
- β-scission of the CS--C(OOH)C• bond generates CH₃S• and C=COOH, with a barrier of 9.0 kcal/mol that is represented by TS234.
- The CSC(OOH)C• radical can also undergo HO₂ elimination to generate CSC=C, with a barrier of 15.3 kcal/mol shown as TS233.

TS241 represents HO₂ molecular elimination from the CSC(OO•)C adduct, generating CSC=C with a barrier of 28.5 kcal/mol. These products are the same as those of the path represented by TS233

The dissociation of CSC(OO•)C into CC=O and CS•=O is illustrated by TS251 and occurs over a barrier of 31.1 kcal mol^{-1}. The peroxy oxygen radical adds to the sulfur. Bond formation between peroxy oxygen and sulfur initiates, along with cleavage of both the sulfur—ethyl carbon and the CH$_3$CH$_2$O—OS•CH3 peroxide bond. Two π bonds are formed: CH3S•=O and CH3CH2=O, and the sulfur–ethyl carbon bond is cleaved.

The intramolecular hydrogen transfer from the primary ethyl carbon to the carbon peroxyl radical (5-member ring) has a 16.3 kcal/mol higher barrier and is 8.2 kcal/mol more endothermic than the 6- member ring TST for the hydrogen transfer from the methyl carbon on the sulfur to the peroxyl radical site.

- Following hydrogen shift, a barrier of 26.0 kcal/mol is required for the y(CSCO)C ring structure formation by C•SC(OOH)C that has a ΔH_{rxn} value of -12.6 kcal/mol.
- The barrier required for the CSy(COC) ring formation from CSC(OOH)C• is 7.8 kcal/mol with the ΔH_{rxn} value of -14.9 kcal/mol.
- The CSC(=O)C ketone formation has a barrier of 36.3 kcal/mol, which is the highest among all the subsequent reactions of CSC(OO•)C.
- The dissociation into CC=O plus CS•=O and the CSC(=O)C ketone formation paths are the more exothermic reactions for CSC(OO•)C, with ΔH_{rxn} value of -53.3 and -36.3 kcal/mol, respectively. Both are more exothermic than reactions leading to a ring structure.
- There are two routes from the initial peroxy radical CSC(OO•)C to CSC=C + HO$_2$: The direct HO$_2$ via TS241 has a 6.1 kcal/mol lower barrier than the TS231path via the ethyl primary radical hydroperoxide CSC(OOH)C•.

The chain branching unimolecular dissociation reaction of CSC(OO•)C → CSC(O•)C + O• has reaction energy of 59.6 kcal/mol. This is some 23 kcal mol^{-1} higher than the highest barriers in the unimolecular reactions presented. The rate constant of this channel is not important at the temperatures and pressures of this study.

The ideal gas-phase thermochemical properties of species calculated from the CBS-QB3 level for the lowest energy geometries are listed in Table 3.

Table 3: Ideal gas-phase thermochemical properties.

Species	$\Delta H_{f\,298}^{\circ}$ [a]	S_{298}° [b]	C_{p300} [b]	C_{p400}	C_{p500}	C_{p600}	C_{p800}	C_{p1000}	C_{p1500}
CSC•C [c]	26.10	66.09	16.86	21.36	25.49	29.04	34.70	39.00	45.83
C•SC(OOH)C [c]	4.30	74.75	21.86	27.41	32.13	36.01	41.93	46.30	53.25
CSC(OO•)C [c]	-4.10	72.18	20.98	26.75	31.86	36.16	42.78	47.63	55.11
CSC(OOH)C• [c]	11.70	75.14	22.02	27.65	32.39	36.26	42.13	46.45	53.33
$CH_3SC(=O)CH_3$ [c]	-49.72	69.91	20.03	24.71	28.87	32.41	37.96	42.09	48.54
CH_2=S	27.60	55.13	9.05	10.29	11.46	12.45	14.02	15.21	17.11
y(CSCO)C	-17.20	69.16	18.92	24.39	29.23	33.27	39.41	43.82	50.50
CSy(CCO)	-11.70	69.23	18.36	23.66	28.31	32.14	37.96	42.16	48.61
$CH_3SCH=CH_2$	15.50	66.16	16.52	20.95	24.83	28.08	33.15	36.94	42.96
$CH_3S•$	29.40	57.69	10.80	12.54	14.07	15.38	17.51	19.16	21.86
$CH_3S•=O$	-17.80	62.48	12.16	14.43	16.42	18.07	20.64	22.54	25.53
O_2	0.00	48.90	7.02	7.20	7.43	7.60	8.00	8.20	8.60
OH	8.89	43.88	7.17	7.08	7.06	7.06	7.15	7.33	7.87
$CH_3CH=O$	-39.60	63.00	13.00	15.50	18.00	20.20	23.80	26.60	30.90
HO_2	2.90	54.60	8.30	8.80	9.40	9.90	10.70	11.30	12.30
$CH_2=CHOOH$	-9.20	73.00	18.40	21.60	24.40	26.70	30.10	32.50	36.20
TS211	14.23	76.32	24.85	31.48	37.04	41.54	48.23	52.94	59.98
TS212	30.34	84.46	29.43	35.40	40.30	44.23	50.13	54.45	61.31
TS213	28.13	90.33	29.07	34.27	38.78	42.51	48.29	52.60	59.46
TS221	32.20	84.38	28.40	34.31	39.42	43.65	50.12	54.78	61.85
TS231	30.49	76.95	23.79	30.19	35.60	39.98	46.51	51.13	58.08
TS232	19.53	91.21	29.87	35.37	40.11	44.00	49.96	54.34	61.28
TS233	26.98	78.05	24.27	29.84	34.52	38.33	44.14	48.42	55.29
TS234	20.68	78.57	24.03	29.55	34.24	38.08	43.95	48.28	55.20
TS241	24.44	78.23	24.36	30.39	35.58	39.87	46.34	50.96	57.95
TS251	27.03	78.98	26.33	32.01	36.94	41.05	47.41	52.08	59.34

[a] Units in kcal mol^{-1}. [b] Units in cal mol^{-1} K^{-1}. [c] Previous work by the same author.

The elementary high-pressure limit Arrhenius rate parameters for the forward and reverse reactions are given in Table 4.

The kinetic rate constants representing •OOCSCC isomerization by intramolecular H transfer reactions have quantum tunneling corrections added to the *ab initio* rate coefficients. The tunneling corrections for the intramolecular hydrogen atom transfer reactions are listed in Table 5.

Table 4: Elementary High-Pressure Rate Parameters by CTST

Reaction	$k = A(T)^n \exp\left(-\dfrac{E_a}{RT}\right)$ $(300 \leq T/K \leq 2000)$		
	$A(s^{-1})$	n	E_a (kcal mol^{-1})
CSC(OO•)C → C•SC(OOH)C	3.8E+7	0.9	1.8E+4
C•SC(OOH)C → CSC(OO•)C	8.5E+8	0.1	9.7E+3
C•SC(OOH)C → CSC-C (cyclic) + OH	1.5E+9	0.4	2.6E+4
C•SC(OOH)C → CH$_2$=S + CC=O + OH	4.5E+12	0.1	2.4E+4
CSC(OO•)C → CSC(=O)C + OH	8.0E+7	1.1	3.6E+4
CSC(OO•)C → CSC(OOH)C•	1.7E+9	0.9	3.4E+4
CSC(OOH)C• → CSC(OO•)C	3.4E+8	0.6	1.8E+4
CSC(OOH)C• → CSC--C (cyclic) + OH	6.2E+8	0.8	7.6E+3
CSC(OOH)C• → CSC=C + HO$_2$	1.6E+11	0.4	1.6E+4
CSC(OOH)C• → CS• + C=COOH	5.3E+9	1.2	8.7E+3
CSC(OO•)C → CSC=C + HO$_2$	1.5E+9	1.1	2.8E+4
CSC(OO•)C → CC=O + CS•=O	1.3E+10	0.3	3.1E+4

Table 5: Quantum tunneling corrections for CSC(OO•)C isomerization by intramolecular H transfer and their reverse reactions.

Temperature (K)	Reactions			
	CSC(OO•)C → C•SC(OOH)C	C•SC(OOH)C → CSC(OO•)C	CSC(OO•)C → CSC(OOH)C•	CSC(OOH)C• → CSC(OO•)C
298	46	26	1150	616
300	46	26	1150	616
350	15.1	10.4	500	300
400	6.1	5.4	46	37
500	3.2	3.1	6.1	6.1
600	2.1	2.1	3.2	3.2
700	1.5	1.5	2.1	2.1
800	1.5	1.5	2.1	2.1
900	1.5	1.5	1.5	1.5
1000	1.2	1.2	1.5	1.5
1100	1.2	1.2	1.5	1.5
1200	1.2	1.2	1.5	1.5

1400	1.2	1.2	1.2	1.2
1500	1.2	1.2	1.2	1.2
1600	1.1	1.1	1.2	1.2
1800	1.1	1.1	1.2	1.2
2000	1.1	1.1	1.2	1.2

Figures 5 and 6 give the plots of reaction rate constants versus 1000/T(K) from 298-2000 K under the pressures of 1 and 4 atm, respectively. 4 atm is chosen as the typical pressure in the diesel engine.

The temperature has a significant effect on the rates of each reaction path under both pressures. Meanwhile, increasing the pressure from 1 to 4 atm does not significantly change the rate of any elementary reaction step under each temperature. Stabilization of and HO_2 elimination from $CH_3SCH(OO\bullet)CH_3$ are both important at room temperature under two pressures. A low temperature enhances $CSC\bullet C$ activation by O_2. The two paths of $CH_3SCH(OO\bullet)CH_3$ isomerization by intramolecular hydrogen shift both become more important at temperatures between 400-1000 K. The rates of molecular dissociations caused by C—S bond β-scission and the formations of $CSC(=O)C$ ketone and $Y(CSCO)C$ and $CSY(CCO)$ ring structures are close to each other above 700 K and are all enhanced by higher temperature. Under a temperature below 400 K, the most and least important reactions besides the isomerization of an O_2 dissociation from $CH_3SCH(OO\bullet)CH_3$ adduct are HO_2 elimination from $CH_3SCH(OO\bullet)CH_3$ and $Y(CSCO)C$ formation by $CH_2\bullet SCH(OOH)CH_3$, respectively.

Figure 5 Rate constant vs. temperature for the CH₃SCH•CH₃ + O₂ reaction system under P=1 atm, where "j" represents the location of the radical site.

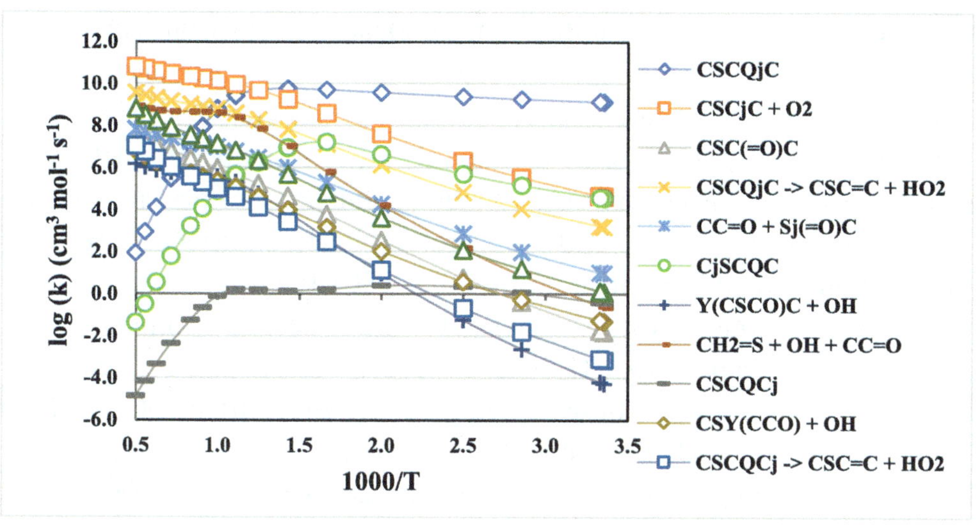

Figure 6 Rate constant vs. temperature for the CH₃SCH•CH₃ + O₂ reaction system under P=4 atm, where "j" represents the location of the radical site.

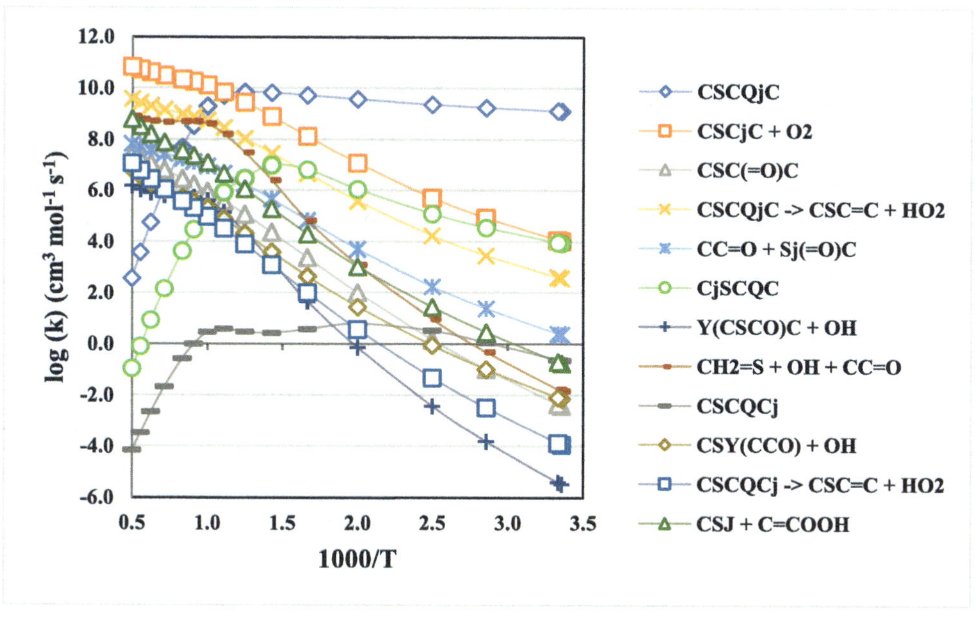

The plots of reaction rate versus pressure under T=298 K and T=700 K are given in Figures 7 and 8, respectively. 700 K is chosen as a typical temperature of sulfuric compound combustion and pyrolysis.

Under T=298 K, the stabilization, dissociation, and isomerization of $CH_3SCH(OO\bullet)CH_3$ adduct are independent of pressure under room temperature. Reaction rates of C—S and O—O bond β-scission, HO_2 elimination, ring structure formation, and CSC(=O)C ketone formation are enhanced under lower pressure. Stabilization to the peroxy radical, intramolecular hydrogen shift generating $CH_2\bullet SCH(OOH)CH_3$, O_2 dissociation, and HO_2 elimination from $CH_2\bullet SCH(OOH)CH_3$ are the most important reactions under all pressures. The rates of intramolecular hydrogen shift leading to $CH_2SCH(OOH)CH_2\bullet$ are about 4-5 orders of magnitude lower than those leading to $CH_2\bullet SCH(OOH)CH_3$. HO_2 elimination from $CH_2SCH(OOH)CH_2\bullet$ and y(CSCO)C formation are among the least important reactions and decrease under higher pressure.

Compared to T=298 K, Figure 8 indicates that pressure has stronger effects on reaction rates under T=700 K. Peroxy radical stabilization remains independent of pressure when the temperature rises from 298 to 700 K. Under T=700 K, intramolecular hydrogen shift leading to $CH_2SCH(OOH)CH_2\bullet$ is the least important reaction and is significantly slower than that leading to $CH_2\bullet SCH(OOH)CH_3$.

Higher pressure reduces all reactions besides the isomerization of an O_2 dissociation from the peroxy radical. Peroxy radical stabilization, O_2 dissociation, and HO_2 elimination from peroxy radical, $CH_2\bullet SCH(OOH)CH_3$ generation, and $CH_2\bullet S$--$CH(OOH)CH_3$ bond β-scission leading to $CH_2=S$, OH, and CC=O are the most important reactions under T=700 K within the pressure range of 0.5-20.0 atm. $CH_3SCH(OO\bullet)CH_3$ dissociation via peroxy radical addition to the sulfur atom is significantly enhanced by increasing the temperature from 298 to 700 K. Meanwhile, decreasing the pressure from 1 to 0.5 atm does not alter any reaction rate under 298 or 700 K. Hence there is no need to pressurize or vacuumize the reaction system to make the CSC$\bullet$C + O_2 reaction system move forward.

Hence Figures 5 and 6 indicate the temperature of 600-800K is recommended for the CH₃SCH•CH₃ + O₂ reaction system to move forward within the pressure range of 1-4 atm.

Figure 7 Rate constant vs. pressure for the CH₃SCH•CH₃ + O₂ reaction system under T= 298 K, where "j" represents the location of the radical site.

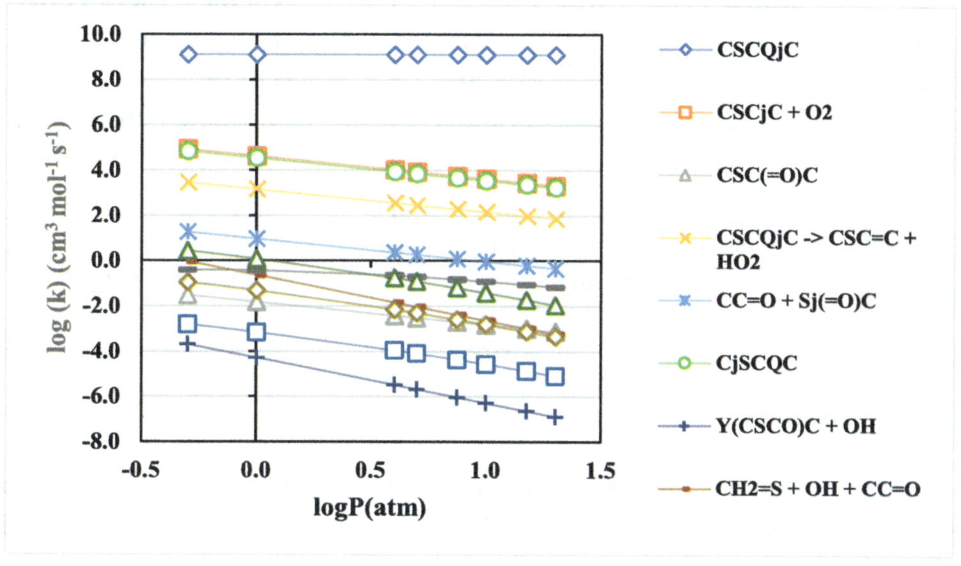

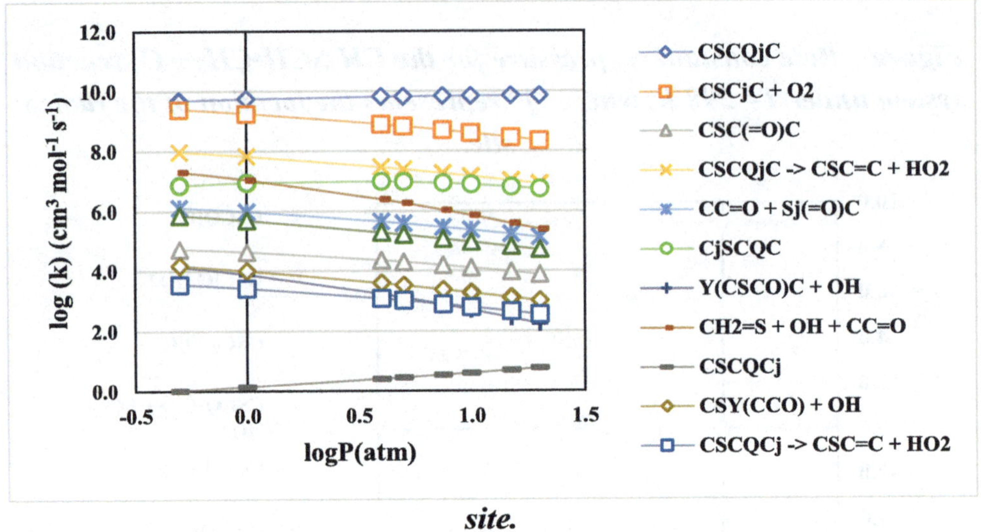

site.

5. Summary

Thermochemical and kinetic studies of the CH₃SCH•CH₃ + O₂ reaction system are presented. The CBS-QB3 composite method of calculation is recommended for the oxidation of organosulfur compounds. The presence of sulfur reduces the chemical activation energy for O₂ association with secondary carbon atoms adjacent to the sulfur moiety by about 7 kcal/mol. The sulfur presents new paths in molecular dissociation by C—S bond β-scission and formations of aldehydes and sulfur – oxygen bond systems.

Several new reaction paths are presented. These involve the addition of the newly formed peroxy oxygen radical to the sulfur with cleavage of the sulfur – carbon bond, which allows the carbon-oxygen carbonyl bond to form.

$$C{\bullet}SC(OOH)C \rightarrow CH_2{=}S + CC{=}O + OH \ (\Delta H_{rxn} = -7.0 \text{ kcal/mol})$$

$$CSC(OOH)C{\bullet} \rightarrow CH_3S{\bullet} + C{=}COOH \quad (\Delta H_{rxn} = 8.5 \text{ kcal/mol})$$

$$CSC(OO{\bullet})C \rightarrow CC{=}O + CS{\bullet}{=}O \qquad (\Delta H_{rxn} = -53.3 \text{ kcal/mol})$$

Rate constants vs. pressure at 300 and 700K and rate constants vs. 1000/T(K) at 1 and 4 atm pressure are presented.

Rising the pressure from 0.5 to 20 atm does not significantly affect any elementary reaction step in the $CH_3SCH \bullet CH_3 + O_2$ reaction system under either 298 or 700 K. Stabilization of $CH_3SCH(OO \bullet)CH_3$ adduct is dominant below 500 K, and minor above 900 K.

The two paths of $CH_3SCH(OO \bullet)CH_3$ adduct isomerization by intramolecular hydrogen transfer are important in the temperature regime 400-800K. All the subsequent reactions of $CH_3SCH(OO \bullet)CH_3$ adduct and its isomers are significantly enhanced by increasing the temperature from 298 to 700 K or higher.

The CBS-QB3, G3MP2B3 composite, and M062x/6-311+G(2d,p) DFT methods are recommended for sulfuric compound oxidation reactions.

The reaction between $CH_3SCH \bullet CH_3$ and O_2 was found to form an energized peroxy adduct $CH_3SCH(OO \bullet)CH_3$ with a calculated well depth of 30.2 kcal/mol at the CBS-QB3 level of theory, under high pressure and low temperature, isomerization and stabilization of the $CH_3SCH(OO \bullet)CH_3$ adduct is of importance, and under atmospheric pressure and at temperatures between above 600 ~ 800 K reactions of the chemically activated peroxy adduct become important relative to stabilization. It has also been found that at temperatures below 500 K, the Stabilization of $CH_3SCH(OO \bullet)CH_3$ adduct is of importance, the temperature of 500-900 K is optimal for intramolecular hydrogen shift, and the isomerization of $CH_3SCH(OO \bullet)CH_3$ adduct, and at a temperature above 800 K, all of the subsequent reaction paths are of importance.

It is concluded that for this oxidation reaction to occur under pressure 1-4 atm, the recommended optimal temperature is between 600-800 K. A new pathway for the $CH_3SCH(OO \bullet)CH_3$ adduct is observed, the attachment of peroxyl oxygen radical to sulfur followed by carbon-sulfur bond dissociation and formation of oxygen-sulfur and oxygen-carbon double bonds

6. Supporting Information Description

Information about the inputs and outputs of the Chemaster code, vibrational frequencies of the transition states under the CBS-QB3 level, and reduced set vibration frequencies under the CBS-QB3 level are given in the Supporting Information.

Enthalpy and Bond dissociation Energy Values for Multi-Fluorinated Ethanol and its Radicals using Gaussian M-062x/6-31+g (d,p) Method at Standard Conditions

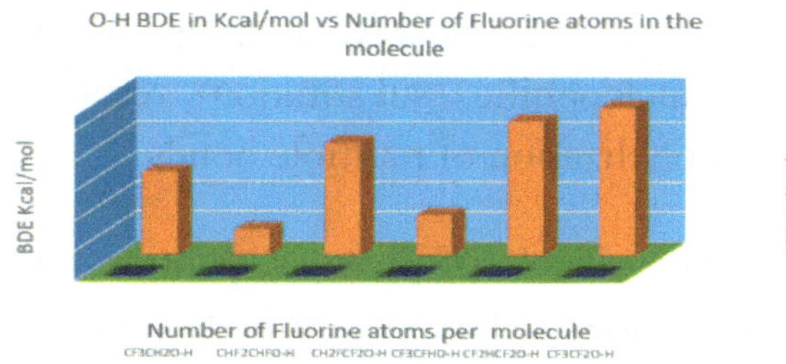

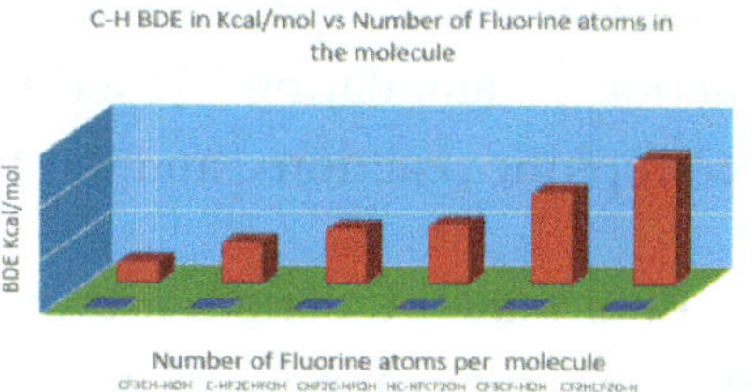

1. Overview

Fluorinated alcohols are used as solvents for proteins, organic compounds, and peptides. They are also known to be used in the organic synthesis industry due to their strong hydrogen bonding character. Halogenated hydrocarbons are mostly synthetically produced, and they don't exist naturally in the environment. Bond dissociation energies values can be used to explain the reactivity and stability of chemical compounds. This chapter presents the methods used to calculate some kinetic and thermodynamic parameters: standard enthalpy of formation and bond dissociation energy values for 18 different fluorinated ethanol, tri-fluorinated ethanol, tetra-fluorinated ethanol, and penta-fluorinated ethanol using the popular ab initio and density functional theory Gaussian M-062x/6-31+g (d,p) method.

Optimized Structures and thermochemical properties of Tri-, Tetra-, and Penta-Fluorinated Ethanol's and its Radicals determined by the Gaussian M-062x/6-31+g (d,p) calculation are presented in this chapter. Tabulated literature values for calculated enthalpies of formation and bond dissociation energies for 18 fluorinated ethanol and some radicals via several series of isodesmic reactions are also included.

2. Introduction

Fluorinated alcohols are known to be used in the organic synthesis industry. They have strong hydrogen bonding donor character, and they are strong nucleophiles that allow organic reactions to occur without the use of a catalyst. Fluorinated alcohols have been used as solvents in epoxidation reactions, annulation reactions, nucleophilic substitution reactions, electrophilic reactions, and the functionalization of multiple bonds.

Fluorinated alcohols are excellent solvents of proteins, peptides, and other organic compounds due to their physicochemical properties. Fluoro-alcohols, Like other alcohols, can alter lipid bilayer properties and stability, and protein function.

Most Halogenated hydrocarbons don't exist naturally in the environment and are synthetically produced. The major source of halogenated hydrocarbons in the atmosphere is agriculture. Crop spraying introduces halogenated hydrocarbons to the environment entering through adsorption and deposition onto airborne particles or directly entering the aquatic system. Some halogenated hydrocarbons, PCBs, furans, and dioxins are byproducts of industrial waste that would unintentionally enter the atmosphere.

The onset temperatures for energetic materials in the calorimetric measurements have been roughly predicted using molecular orbital calculations of bond dissociation energies. The stability and reactivity of chemical compounds can be explained using bond dissociation energies values. Standard enthalpies of formation estimated using semi-empirical MO calculations, the MOPAC-PM7 package has been used previously to derive bond dissociation energy values for chemical compounds.

Bond dissociation energies (BDEs) values for some inorganic compounds, lanthanide selenides, and sulfides, were measured in 2021 using Resonant two-photon ionization spectroscopy. The predissociation thresholds were found to be the BDE values for these molecules. The 0 K gaseous heat of formation, $\Delta_f H \circ K$ for each molecule, was also reported using this method.

The amount of energy used to break a mole of the covalently bonded gas molecule to a pair of radicals is the bond dissociation energy. The units used for the bond dissociation energy are commonly kJ/mol. Covalent bonds can be broken heterolytically or homolytically. The heterolytic breaking of a covalent bond would result in the pair of electrons going to only one atom, either A or B.

$C{-}D{\rightarrow}C^+ + D^{:-}$ or $C{-}D{\rightarrow}C^{:-} + D^+$. The homolytic breaking of a covalent bond, on the other hand, would result in one electron staying with each atom, $\cdot A{-}B{\rightarrow}A{\bullet}{+}B{\bullet}$. Bond dissociation energy can be calculated for molecules as the difference in enthalpy of the formation of products and reactants. Bond dissociation energy is a state function, as it doesn't depend on the mechanism or pathway of how bonds form or break. The energy of chemical reactions can be assessed using values for the bond dissociation energy. There are some systematic trends for the bond dissociation values; bond dissociation energy varies with hybridization. For example, sp3 hybridized carbons in hydrocarbons have smaller bond dissociation values compared to sp2 hybridized carbons. The longer and weaker sp3 hybridized bond is easier to break compared to the shorter and stronger sp2 and sp hybridized bonds (double and triple bonds). Among sp3 hybridized bonds, bond dissociation values depend on their position, whether it's on a primary, secondary, or tertiary carbon. Methane has the strongest C-H bond with the highest bond dissociation values, following C-H bonds on primary carbons, following C-H bonds on secondary carbons, and following C-H bonds on tertiary carbons.

Energetics of chemical processes can be assessed using bond dissociation energy values. Hess's Law has been used in the past and is currently being used to estimate reaction enthalpies by combining bond dissociation energies of bonds formed and bond dissociation energies of bonds broken.

The energy change when forming a mole of compound from its component elements is called the enthalpy of formation, ΔH_f. If heat is released when the elements combine to form the compound, enthalpy of formation would have a negative sign. If heat is absorbed when the elements combine to form the compound, the enthalpy of formation would have a positive sign Values of the Enthalpy of formation are dependent upon temperature, pressure, and physical states of reactants and products in the chemical reaction. Standard enthalpy of formation ($\Delta H°_f$) is the enthalpy of formation at standard conditions; 1atm pressure, 25°C, and 1M aqueous solution concentration. Any element in its most stable form has a standard enthalpy of formation has a value of zero. Tabulated enthalpy of formation values can be used to calculate the standard enthalpy of any reaction whose standard enthalpy of formation values are well known,

$$\Delta H^o_{rxn} = \sum m\Delta H^o_f\,(products) - \sum n\Delta H^o_f\,(reactants)$$

The heat involved in chemical or physical change at constant temperature and pressure is the enthalpy of reaction (H), and it's a thermodynamic quantity, q=ΔH.

Molecules, ions, or atoms containing at least an unpaired electron in the valance shell are called free radicals. Free radicals are chemically reactive, unstable, and mostly short-lived. Heat, electrolysis, electrical discharge, and ionizing radiation can generate free radicals. Free radicals are intermediates in many chemical reactions. Free radicals are important in atmospheric chemistry, combustion, plasma chemistry, polymerization, and biochemistry, and they are important in many chemical reactions.

In 2016 Hang Wang studied the thermodynamic properties of fluorinated methanol using CBS-QB3, M06-2X, WB97X, W1U, M06, B3LYP, CBS-APNO, and G4 Calculations. A small standard deviation suggests good error cancellation of work reactions and accuracy. Small values for standard deviations were obtained in calculations using the M06-2x/6-31+g (d,p) Gaussian method; it is an accurate method to calculate the Enthalpy of fluorinated alcohols; it shows the second smallest standard deviation after the CBS-QB3 method of calculation. The enthalpy of fluorinated methanol was studied in the past.

Halogenated compounds have low reactivity, are highly stable, and are used in industry. They are of concern to the environment due to their persistence in the environment and their widespread use. Their thermochemical properties must be studied in order to understand the reduction and oxidation reactions involving these molecules.

3. Experimental and Data

3.1 Computational Method

The Global-hybrid meta-GGA density functional approximation, GGA, DFT Gaussian M-062x/6-31+g (d,p) method of calculation has been used to initially analyze frequencies, optimized structures, and thermo energies of the molecules studied. In the GGA, generalized gradient approximation, the density functional depends on the down and up spin densities and the reduced gradient. In the meta GGA, the function also depends on the up and down spin kinetic energy densities. A hybrid GGA is a combination of GGA with the Hartree-Fock exchange. The hybrid meta GGA is a combination of meta GGA with the Hartree-Fock exchange.

3.2 Isodesmic and Isogyric Reaction

A series of Isodesmic Reactions and composite calculations were employed to calculate the enthalpy of the formation of tri-, terta-, and penta-fluorinated ethanol. Calculations were performed using Gaussian 16 program. All reported calculations **of** enthalpy of formation are for standard conditions of 1 atm pressure and 298K. The Gaussian M-062x/6-31+g (d,p) level of calculation has been used for this study as this method of calculation was successfully employed in the past when applied to fluoro hydrocarbons 6 with small reported standard deviation values.

The standard deviation is calculated using the following formula·

$$\sigma = \sqrt{\frac{1}{N}\sum_{i=1}^{N}(x_i - \mu)^2}$$

Where, X_i is the mean; the average of the numbers, μ is the actual numbers to be calculated the standard deviation of, and

$$\frac{1}{N} \sum_{i=1}^{N} (x_i - \mu)^2$$

is the variance.

The Gaussian M-062x/6-31+g (d,p) method of calculation used to calculate the enthalpy of formation of tri-, terta-, and penta-fluorinated ethanol's and its radicals are presented. Work reactions and reference species used to calculate the enthalpy of formation of fluorinated ethanol using this method is also presented. The number of each type of bond must be conserved in each isodesmic reaction to cancel any systematic error in the molecular orbital calculations using this method. By the careful choice of the isodesmic reactions, all enthalpies of formation calculations are allowed to accuracies close to experimental values. The ΔH_{f298K}° values of all reference species but the fluoroethanols are known; the $\Delta H_{f298^{\circ}K}$ of the target species fluoro-ethanols is obtained from this data, and the calculated ΔH_{rxn}, 298 °. $\Delta H_{f298^{\circ}K}$ calculated using two different reference molecules are within $\pm$ (0 to 0.60 Kcal mol^{-1}).

3.3 Reference Species

Table 1 lists the Standard enthalpy of formation for the reference species used in isodesmic reactions with their uncertainties. Table 2 provides all calculated standard enthalpy of formation values, $\Delta_f H^{\circ}{}_{(298)}$ for tri-, terta-, and penta-fluorinated ethanol and its radicals.

Table 1. Reference Species in the Isodesmic Reactions Standard Enthalpy of Formation Values (kcal mol−1)

Species	$\Delta_f H^{O}{}_{(298)}$	Species	$\Delta_f H^{O}{}_{(298)}$

CH_3F	-56.54 ± 0.07[a] -56.62 ± 0.48[h]	CH_3OOH CH_3CH_2OOH	-30.96 ± 0.67[b] -38.94 ± 0.81[b]
CH_3CH_2F	-65.42 ± 1.11[a]	$CH_3CH_2CH_2OOH$	-44.03 ± 0.67[b]
$CH_3CH_2CH_2F$	-70.24 ± 1.30[a]	$CH_3OO^\bullet$	2.37 ± 1.24[b]
CH_2F_2	-108.07 ± 1.46[a] -107.67 ± 0.48[h]	$CH_3CH_2OO^\bullet$ $CH_3CH_2CH_2OO^\bullet$	-6.19 ± 0.92[b] -11.35 ± 1.24[b]
CH_3CHF_2	-120.87 ± 1.62[a]	CH_4	-17.81 ± 0.01[c]
$CH_3CH_2CHF_2$	-125.82 ± 1.65[a]	CH_3CH_3	-20.05 ± 0.04[c]
CHF_3	-166.71 ± 1.97[h] -166.09 ± 0.48[h]	$CH_3CH_2CH_3$ $CH_3CH_2CH_2CH_3$	-25.01 ± 0.06[i] -30.07 ± 0.08[i]
CH_3CF_3	-180.51 ± 2.05[a]	$CH_3O^\bullet$	5.15 ± 0.08[c]
$CH_3CH_2CF_3$	-185.48 ± 2.15[a]	$CH_3CH_2O^\bullet$	-3.01[d]
$CH_3^\bullet$	34.98 ± 0.02[c]	OH	8.96 ± 0.01[c]
$CH_3CH_2^\bullet$	28.65 ± 0.07[c]	CH_3OH	-47.97 ± 0.04[c]
$CH_3CH_2CH_2^\bullet$	24.21 ± 0.24[gj] 24.18[i]	CH_3CH_2OH	-56.07 ± 0.05[i]
H	52.10[c]	$HOO^\bullet$	2.94[cj]
O	59.57[c]	$HOOH$	-32.39 ± 0.04[fj] -32.37[i]
CF_4 [13]	-223.15	CHF_2CHF_2 [13]	-212.13
$CF_3CH_2CHF_2$ [13]	-286.18	CF_3CHF_2 [13]	-267.79

3.4 Standard Enthalpies

Isodesmic work reactions from M-062x/6-31+g (d,p) method of calculation utilized to perform the calculation of standard enthalpy of formations for tri-, tertra-, and penta- fluorinated ethanol's and some radicals are presented table 2. Reference species and their standard enthalpy of formation, along with their uncertainties, have been used in all isodesmic work reactions, table 1. The calculated sum of thermal enthalpies using the Gaussian M06-2x/6-31+g (d,p) Level of Theory for all target fluorinated ethanol's, their radicals, and reference species has also been used. The standard enthalpy of formation in kcal mol^{-1} for all reference species is listed in table 1. Standard deviation values are listed for all calculated Gaussian M06-2x/6-31+g (d,p) standard enthalpy of formations values. The calculated Gaussian M06-2x/6-31+g (d,p) standard enthalpy of formation for tri-, tertra-, and penta-fluorinated ethanol and some radicals are listed table2.

Table 2. Isodesmic reactions used in calculating the standard enthalpy of formation, ΔH°$_{Rxn}$ for tri-, terta-, and penta-fluorinated ethanol's and its radicals using the Gaussian M06-2x/6-31+g (d,p) Level of Theory

Isodesmic Reactions Target Specie	ΔH°$_{Rxn}$ (298) Hartrees	ΔH°$_{Rxn (298)}$ Kcal/mole[1]	Δ$_f$H°$_{(298)}$ kcal mol^{-1}	Error kcal mol^{-1}
CF3CH2OH + CH4 = CH3CH2OH + CHF3 -452.696144 -40.447961 -154.926666 -338.204669 -17.81 -56.21 -166.71	0.01277	8.013303	-213.1	±2.03
CF3CH2OH + CH3CH3 = CH3CH2OH + CH3CF3 -452.696144 -79.717768 -154.926666 - 377.492814 -20.05 -56.21 -180.51	-0.00557	-3.49398	-213.2	±2.14
Reported Δ$_f$H° (298) kcal mol^{-1}			-213.15 ±2.09	
Standard Deviation over rxns 0.05 kcal mol^{-1}				
CF3CH•OH + CH3CH3 = CH3CH2O• + CH3CF3 -452.048269 -79.717768 -154.268107 -377.492814 -20.05 -3.01 -180.51	0.005116	3.210341	-166.7	± 2.09
CF3CH•OH + CH4 = CH3CH2O• + CHF3 -452.048269 -40.447961 -154.268107 -338.204669 -17.81 -3.01 -166.71	0.023454	14.71762	-166.6	±1.98
Reported Δ$_f$H° (298) kcal mol^{-1}			-166.65±2.04	
Standard Deviation over rxns 0.05 kcal mol^{-1}				
CF3CH2O• +CH3CH2CH3= CH3CH2O• + CH3CH2CF3 -452.023032 -118.99092 -154.268107 - 416.766483 -25.02 -3.01 -185.48	-0.02064	-12.9537	-150.5	±2.21
CF3CH2O• + CH4 = CH3CH2O• + CHF3 -452.023032 -40.447961 -154.268107 - 338.204669 -17.81 -3.01 -166.71	-0.00178	-1.11885	-150.8	±1.98
Reported Δ$_f$H° (298) kcal mol^{-1}			-150.70±2.09	
Standard Deviation over rxns 0.16 kcal mol^{-1}				
CHF2CHFOH + CH3CH3 = CH3CH2OH + CH3CF3	-0.0152	-9.5369	-207.1	±2.14

-452.686514 -79.717768 -154.926666 -377.492814 -20.05 -56.21 -180.51				
CHF2CHFOH + CH4 = CH3CH2OH + CHF3 -452.686514 -40.447961 -154.926666 - 338.204669 -17.81 -56.21 -166.71	0.00314	1.970381	-207.1	±2.03
Reported $\Delta_f H°$ (298) kcal mol^{-1}			-207.10±2.09	
Standard Deviation over rxns 0.00 kcal mol^{-1}				
C•F2CHFOH + CH4 = CH3CH2O• +CHF3 -452.032761 -40.447961 -154.268107 - 338.204669 -17.81 -3.01 -166.71	0.007946	4.986194	-156.9	±1.98
C•F2CHFOH + CH3CH3 = CH3CH2O• + CH3CF3 -452.032761 -79.717768 -154.268107 -377.492814 -20.05 -3.01 -180.51	-0.01039	-6.52108	-156.9	±2.09
Reported $\Delta_f H°$ (298) kcal mol^{-1}			-156.90 ±2.04	
Standard Deviation over rxns 0.00 kcal mol^{-1}				
CHF2C•FOH + CH3CH3 = CH3CH2O• + CH3CF3 -452.028711 -79.717768 -154.268107 -377.492814 -20.05 -5.01 -180.51	-0.01444	-9.0625	-154.4	±2,09
CHF2C•FOH + CH3CH2CH3 = CH3CH2O• + CH3CH2CF3 -452.028711 -118.990915 -154.268107 - 416.766483 -25.02 -5.01 -185.48	-0.01496	-9.39006	-154.1	±2.21
Reported $\Delta_f H°$ (298) kcal mol^{-1}			-154.30±2.15	
Standard Deviation over rxns 0.16 kcal mol^{-1}				
CHF2CHFO• + CH4 = CH3CH2O• + CHF3 -452.019492 -40.447961 -154.268107 -338.204669 -17.81 -3.01 -166.71	-0.00532	-3.34024	-148.6	±1.98
CHF2CHFO• + CH3CH2CH3 = CH3CH2O• + CH3CH2CF3	-0.02418	-15.1751	-148.3	±2.21

-452.019492 -118.99092-154.268107 - 416.766483 -25.02 -3.01 -185.48				
Reported $\Delta_f H^o$ (298) kcal mol^{-1}			-148.50±2.09	
Standard Deviation over rxns 0.16 kcal mol^{-1}				
CH2FCF2OH + CH4 = CH3CH2OH + CHF3 -452.696009 -40.447961 -154.926666 -338.204669 -17.81 -56.21 -166.71	0.012635	7.928589	-213.0	±2.03
CH2FCF2OH +CH3CH3 = CH3CH2OH + CH3CF3 -452.696009 -79.717768 -154.926666 - 377.492814 -20.05 -56.21 -180.51	-0.0057	-3.57869	-213.1	±2.14
Reported $\Delta_f H^o$ (298) kcal mol^{-1}			-213.05±2.09	
Standard Deviation over rxns 0.05 kcal mol^{-1}				
C•HFCF2OH + CH4 = CH3CH2O•+ CHF3 -452.036986 -40.447961 -154.268107 -338.204669 -17.81 -3.01 -166.71	0.012171	7.637424	-159.5	±1.98
C•HFCF2OH + CH3CH3 = CH3CH2O•+ CH3CF3 -452.036986 -79.717768 -154.268107 -377.492814 -20.05 -3.01 -180.51	-0.00617	-3.86985	-159.6	±2.09
Reported $\Delta_f H^o$ (298) kcal mol^{-1}			-159.55±2.04	
Standard Deviation over rxns 0.08 kcal mol^{-1}				
CH2FCF2O• + CH4 = CH3CH2O• +CHF3 -452.019556 -40.447961 -154.268107 -338.204669 -17.81 -3.01 -166.71	-0.00526	-3.30008	-148.6	±1.98
CH2FCF2O• + CH3CH3 = CH3CH2O•+ CH3CF3 -452.019556 -79.717768 -154.268107 -377.492814 -20.05 -3.01 -180.51	-0.0236	-14.8074	-148.7	±2.09
Reported $\Delta_f H^o$ (298) kcal mol^{-1}			-148.65±2.04	
Standard Deviation over rxns 0.05 kcal mol^{-1}				
CF3CFHOH + CH4 = CH3CH2OH+ CF4	0.004717	2.959965	-264.5	±0.10

-551.950082 -40.447961 -154.926666 -437.46666 -17.81 -56.21 -223.15				
CF3CFHOH + CH3CH3 = CH3CH2OH + CHF2CHF2 -551.950082 -79.717768 -154.926666 - 476.716001 -20.05 -56.21 -212.13	0.025183	15.80258	-264.1	±0.09
Reported $\Delta_f H^o$ (298) kcal mol^{-1}			-264.30±0.10	
Standard Deviation over rxns 0.20 kcal mol^{-1}				
CF3C•FOH + CH4 = CH3CH2O•+ CF4 -551.293882 -40.447961 -154.268107 -437.46666 -17.81 -3.01 -223.15	0.007076	4.440261	-212.8	±0.10
CF3C•FOH + CH3CH3 = CH3CH2O• + CHF2CHF2 -551.293882 - 79.717768 -154.268107 -476.716001 -20.05 -3.01 -212.13	0.027542	17.28288	-212.4	±0.40
Reported $\Delta_f H^o$ (298) kcal mol^{-1}			-212.60±0.30	
Standard Deviation over rxns 0.20 kcal mol^{-1}				
CF3CFHO• + CH4 = CH3CH2O•+ CF4 -551.281264 -40.447961 -154.268107 -437.46666 -17.81 -3.01 -223.15	-0.00554	-3.47766	-204.9	±0.10
CF3CFHO• + CH3CH3 = CH3CH2O• + CHF2CHF2 -551.281264 -79.717768 -154.268107 -476.716001 -20.05 -3.01 -212.13	0.014924	9.364959	-204.5	±0.40
Reported $\Delta_f H^o$ (298) kcal mol^{-1}			-204.70±0.30	
Standard Deviation over rxns 0.20 kcal mol^{-1}				
CF2HCF2OH + CH4 = CH3CH2OH+ CF4	0.002188	1.372992	-262.9	±0.10

-551.947553 - 40.447961 -154.926666 -437.46666 -17.81 -56.21 -223.15				
CF2HCF2OH + CH3CH3 = CH3CH2OH+ CHF2CHF2 -551.947553 -79.717768 -154.926666 -476.716001 -20.05 -56.21 -212.13	0.022654	14.21561	-262.5	±0.10
Reported $\Delta_f H^o$ (298) kcal mol^{-1}			-262.70±0.10	
Standard Deviation over rxns 0.20 kcal mol^{-1}				
C•F2CF2OH + CH4 = CH3CH2O• + CF4 -551.286788 -40.447961 -154.268107 -437.46666 -17.81 -3.01 -223.15	-1.8E-05	-0.0113	-208.3	±0.10
C•F2CF2OH + CH3CH3 = CH3CH2O• + CHF2CHF2 -551.286788 -79.717768 -154.268107 - 476.716001 -20.05 -3.01 -212.13	0.020448	12.83132	-207.9	±0.40
Reported $\Delta_f H^o$ (298) kcal mol^{-1}			-208.10±0.30	
Standard Deviation over rxns 0.20 kcal mol^{-1}				
CF2HCF2O• + CH4 = CH3CH2O• + CF4 -551.26873-40.447961 -154.268107 -437.46666 -17.81 -3.01 -223.15	-0.01808	-11.3429	-197.0	±0.10
CF2HCF2O• + CH3CH3 = CH3CH2O• + CHF2CHF2 -551.26873 -79.717768 -154.268107 -476.716001 -20.05 -3.01 -212.13	0.00239	1.499749	-196.6	±0.40
Reported $\Delta_f H^o$ (298) kcal mol^{-1}			-196.80±0.30	
Standard Deviation over rxns 0.20 kcal mol^{-1}				
CF3CF2OH + CH3CH2CH3= CH3CH2OH+CF3CH2CHF2 -651.207522 -118.99092-154.926666 -615.260935	0.010836	6.799698	-324.2	±0.10

-25.02 -56.21 -286.18				
CF3CF2OH + CH3CH3 = CH3CH2OH+ CF3CHF2 -651.207522 -79.717768 -154.926666 -575.976217 -20.05 -56.21 -267.79	0.022407	14.06062	-318.0	±1.70
CF3CF2OH + CH4 = CF3OH+ CH3CHF2 -651.207522 -40.447961 -413.437575-278.225303 -17.81 -218.11 -120.87	-0.00739	-4.64044	-316.5	±1.70
Reported Δ_fH° (298) kcal mol^{-1}			-319.60±1.20	
Standard Deviation over rxns 3.30 kcal mol^{-1}				
CF3CF2O• + CH4 = CHF2O• + CH3CF3 -650.527081 -40.447961 -313.498636 -377.492814 -17.81 -97.82 -180.51	-0.01641	-10.2962	-250.2	±2.06
CF3CF2O• + CH3CH2CH3 = CH3CH2O•+CF3CH2CHF2 -650.527081 -118.990915 -154.268107 - 615.260935 -25.02 -3.01 -286.18	-0.01105	-6.93148	-257.2	±0.10
CF3CF2O• + CH3CH3 = CH3CH2O•+ CF3CHF2 -650.527081 -79.717768 -154.268107 -575.976217 -20.05 -3.01 -267.79	0.000525	0.329443	-251.1	±1.70
Reported Δ_fH° (298) kcal mol^{-1}			-252.80±1.30	
Standard Deviation over rxns 3.11 kcal mol^{-1}				

Hartrees, kcal mole^{-1}
*SD Standard Deviation kcal mol^{-1}
Errors reported avg of sum of uncertainties in rxn's reference species

3.5 Bond Energies

The calculated bond dissociation energy values in kcal/ mol for Methyl C-H, Ethyl C-H, and hydroxyl O-H bonds are listed in table 3.

The difference in bond dissociation energies of products and reactants for hemolysis is called the bond dissociation energy. The bond dissociation energy values don't depend on the pathway by which it occurs; it doesn't depend on how bonds are formed or on how bonds break. Therefore, bond dissociation energies are state functions. Energetics of chemical processes can be assessed using bond dissociation energy values.

Table 3. Bond Dissociation Energy (BDE's) of Tri-, Tetra-, and Penta-Fluorinated Ethanol's Using Gaussian M-062x/6-31+g (d,p) method of calculation.

Reactions	Bond Dissociation Energy[a] (Kcal mol^{-1})	* Error Kcal mol^{-1}
CF3CH-HOH CF3CH-HOH = H• + CF3CH• OH -213.15 ±2.09 **52.10** -166.65±2.04	**98.60± 2.07**	**± 2.07**
CF3CH2O-H CF3CH2O-H = H• + CF3CH2O• -213.15 ±2.09 **52.10** -150.70±2.09	**114.55±2.09**	**±2.09**
C-HF2CHFOH C-HF2CHFOH = H• + C• F2CHFOH -207.10±2.09 **52.10** -156.90 ±2.04	**102.30±2.07**	**±2.07**
CHF2C-HFOH CHF2C-HFOH = H• + CHF2C• FOH -207.10±2.09 **52.10** -154.30±2.15	**104.90±2.10**	**±2.10**

<u>CHF2CHFO-H</u> CHF2CHFO-H = H• + CHF2CHFO• -207.10±2.09 **52.10** -148.50±2.09	**110.70±2.09**	**±2.09**
<u>HC-HFCF2OH</u> HC-HFCF2OH = H• + C• HFCF2OH -213.05±2.09 **52.10** -159.55±2.04		
<u>CH2FCF2O-H</u> CH2FCF2O-H= H• + CH2FCF2O• -213.05±2.09 **52.10** -148.65±2.04	**105.60±2.07**	**±2.07**
	116.50±2.07	**±2.07**
<u>CF3CF-HOH</u> CF3CF-HOH= H• + CF3CF• OH -264.30±0.10 **52.10** -212.60±0.30		
	103.80±0.20	**±0.20**
<u>CF3CFHO-H</u> CF3CFHO-H= H• + CF3CFHO• -264.30±0.10 **52.10** -204.70±0.30		
	111.70±0.20	**±0.20**
<u>CF2-HCF2OH</u> CF2-HCF2OH= H• + CF2• CF2OH -262.70±0.10 **52.10** -208.10±0.30		
	106.70±0.20	**±0.20**
<u>CF2HCF2O-H</u> CF2HCF2O-H= H• + CF2HCF2O• -262.70±0.10 **52.10** -196.80±0.30		
	118.00±0.20	**±0.20**
<u>CF3CF2O-H</u> CF3CF2O-H= H• + **CF3CF2O•** -319.60±1.20 **52.10** -252.80±1.30		

	118.90±1.30	±1.30

*Errors reported avg of sum of uncertainties in rxn's reference specie

5. Optimized Structures

The GaussView software has been utilized along with gaussian output files to provide a picture of the Gaussian M06-2x/6-31+g (d,p) optimized structure for each molecule in this chapter, table 4.

Table 4. Optimized geometries for target tri-, tetra-, and penta- fluorinated ethanol's and their related radicals calculated by Gaussian M06-2x/6-31+g (d,p) level of theory.

CF₃CH₂OH

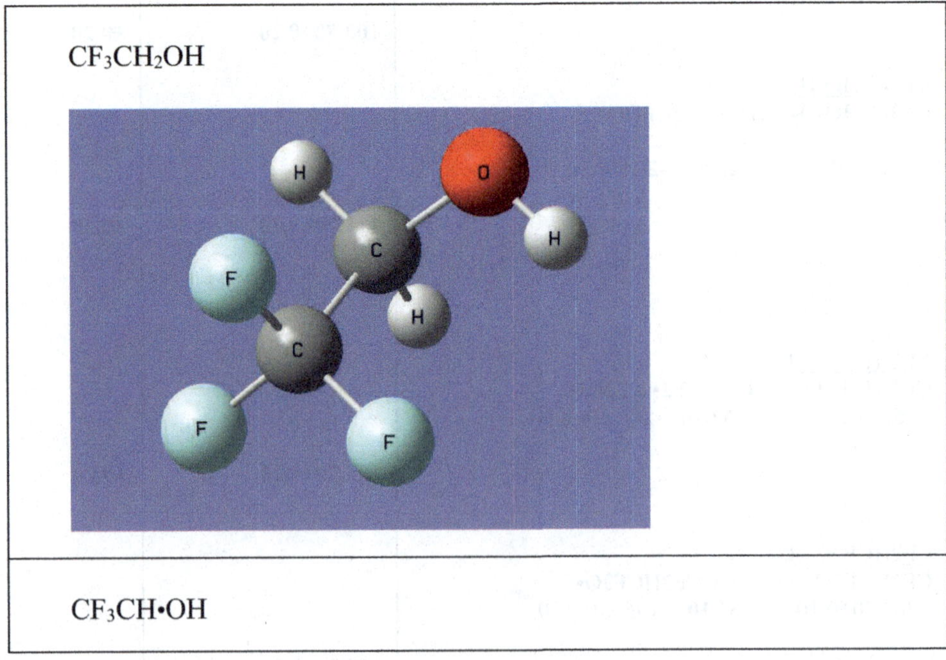

CF₃CH•OH

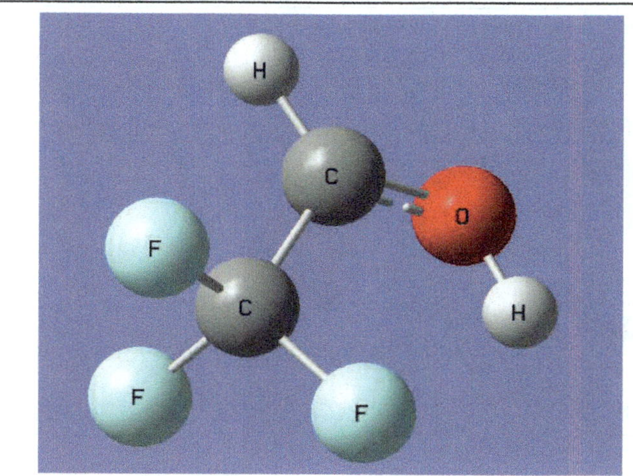

CF$_3$CH$_2$O•

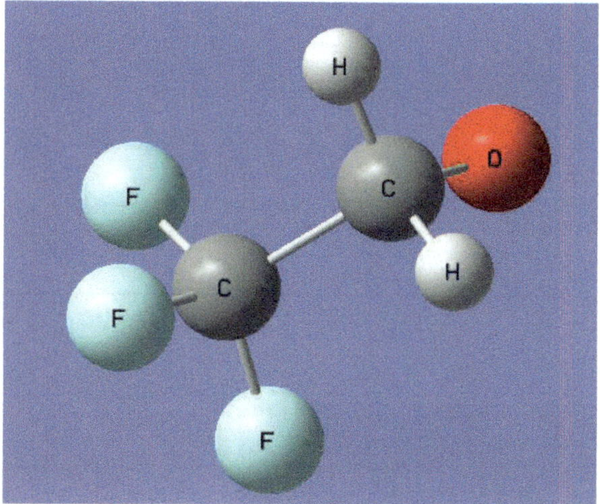

CHF$_2$CHFOH

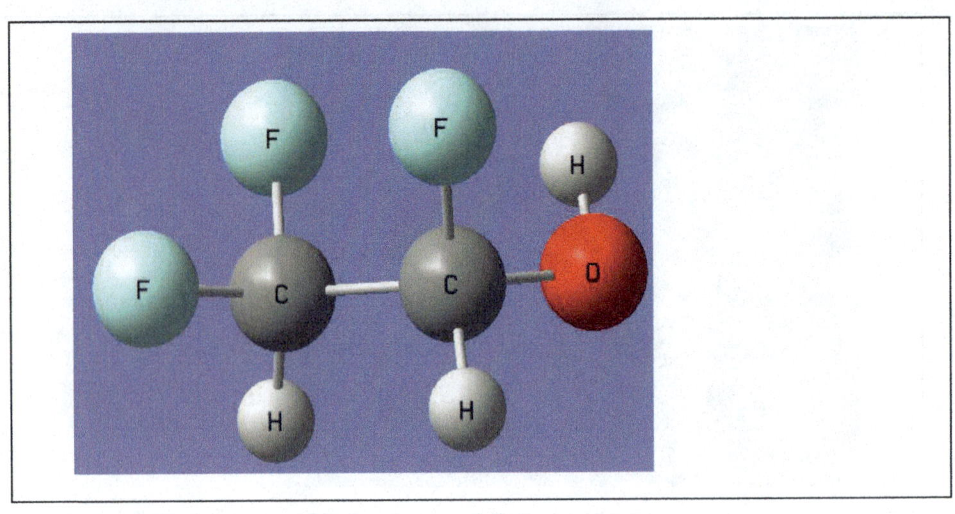

C•F$_2$CHFOH

CHF$_2$C·FOH

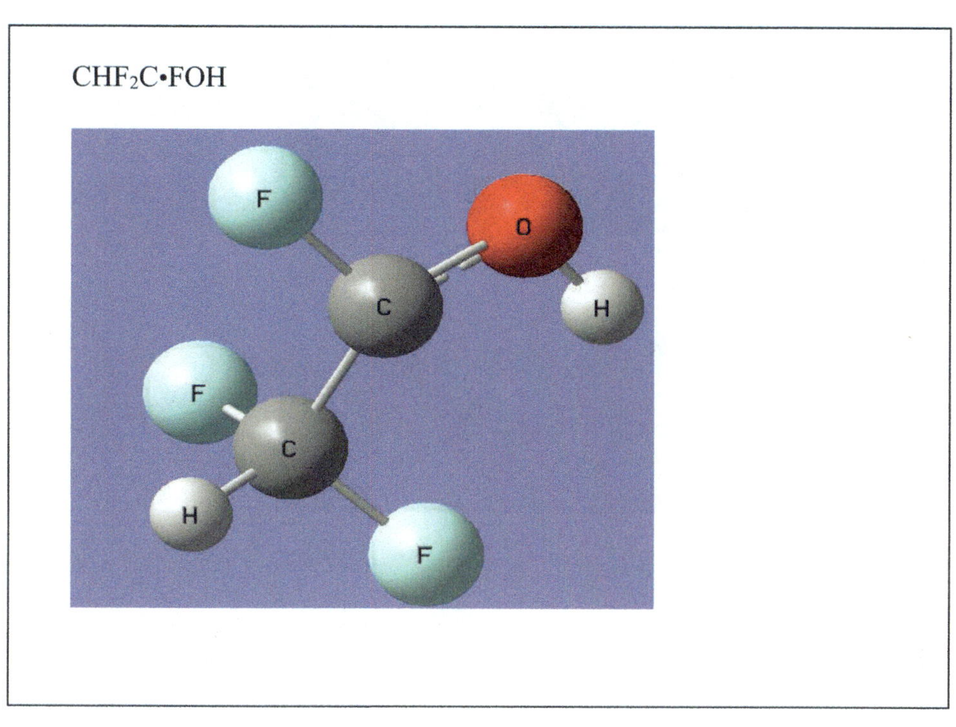

CHF$_2$CHFO·

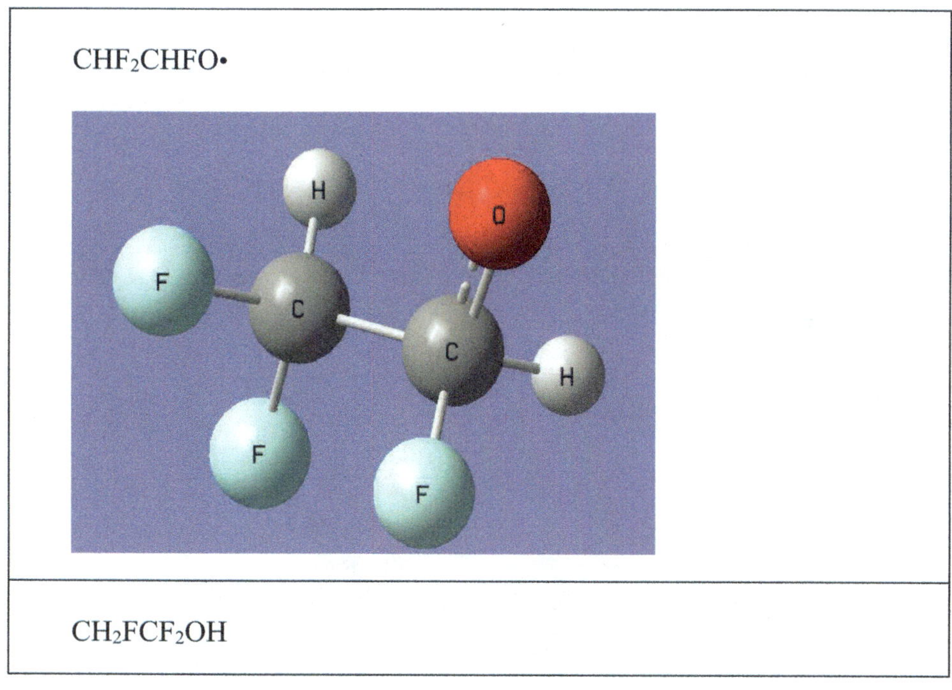

CH$_2$FCF$_2$OH

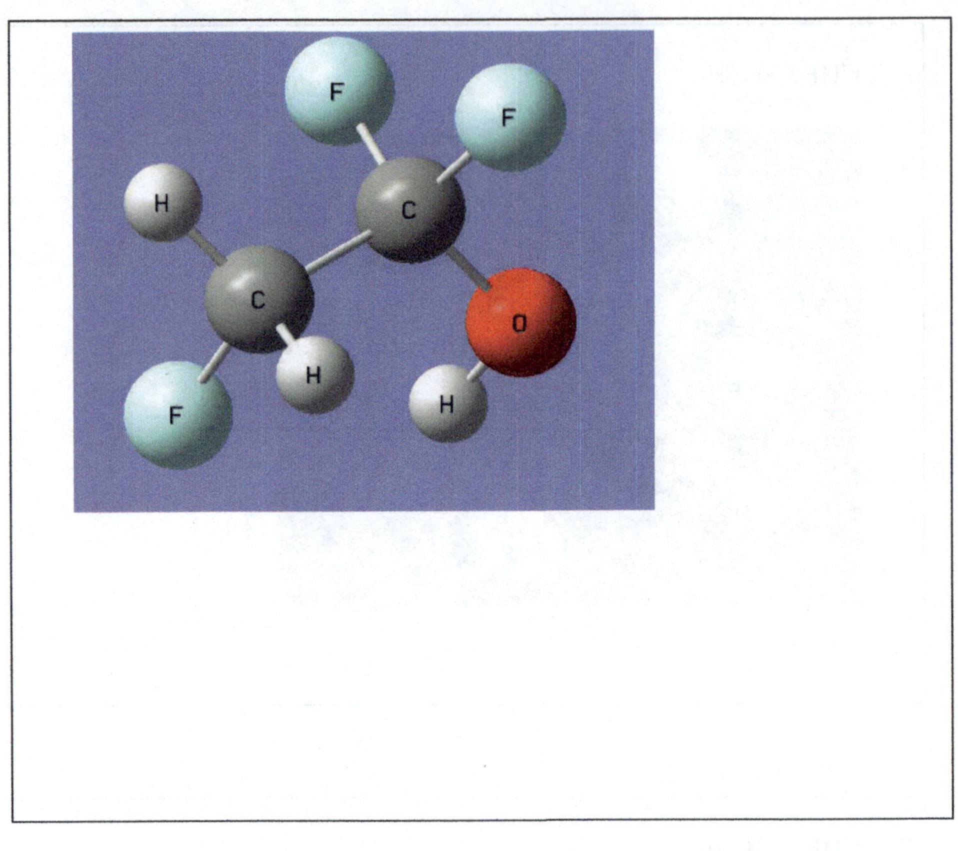

C·HFCF$_2$OH

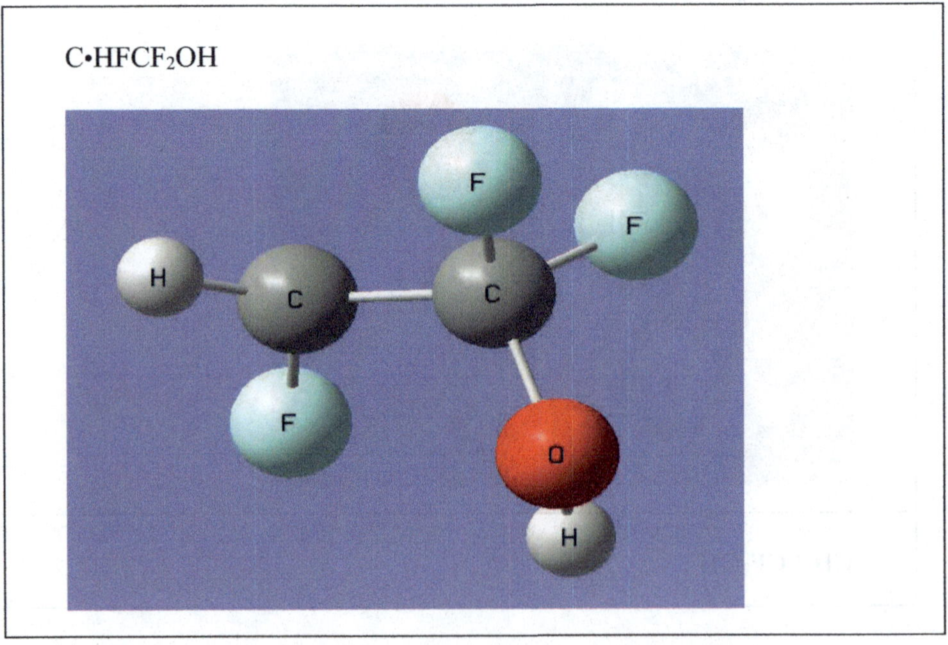

CH$_2$FCF$_2$O•

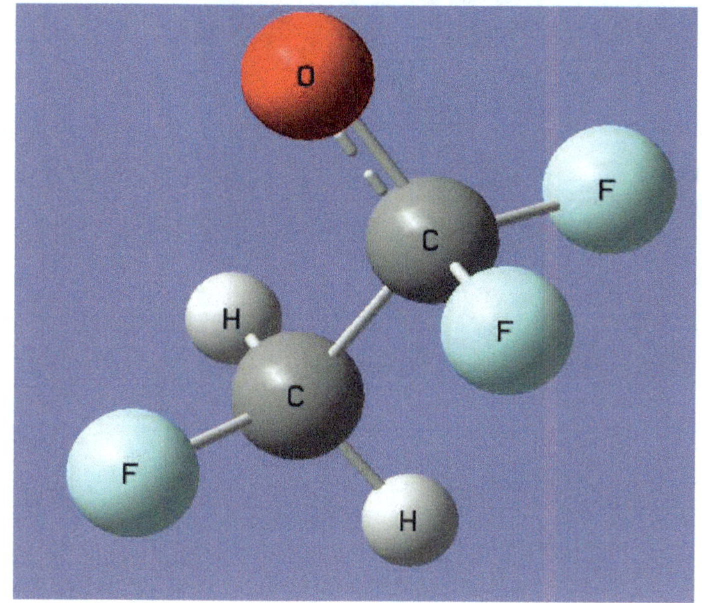

CF$_3$CFHOH

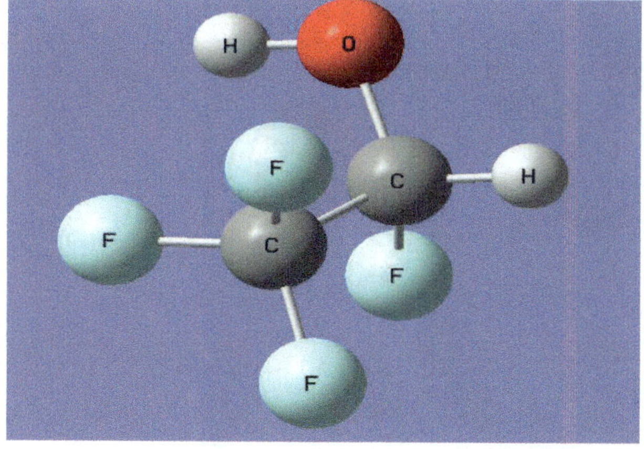

$CF_3CF \cdot OH$

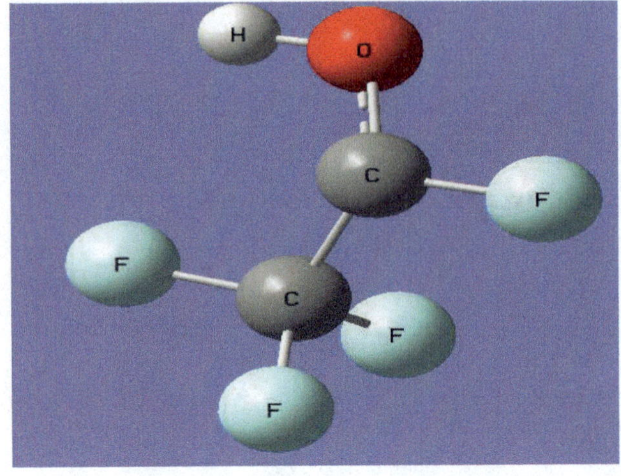

$CF_3CFHO \cdot$

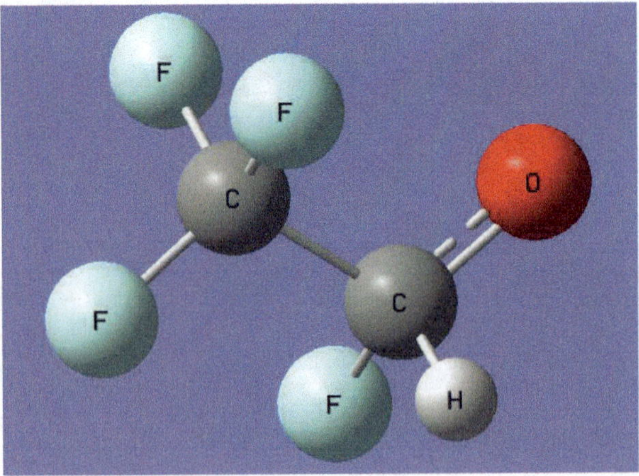

CF_2HCF_2OH

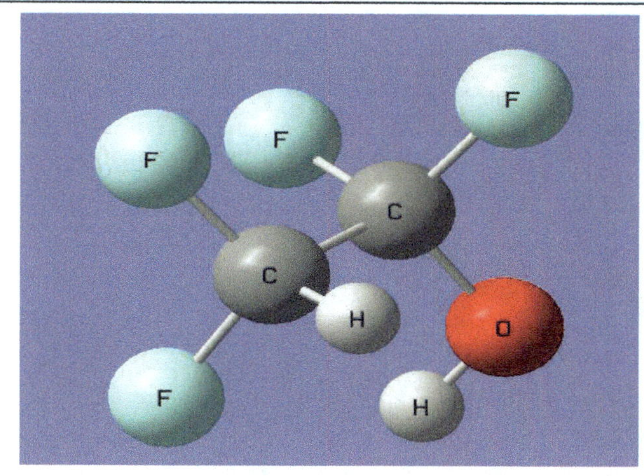

C•F$_2$CF$_2$OH

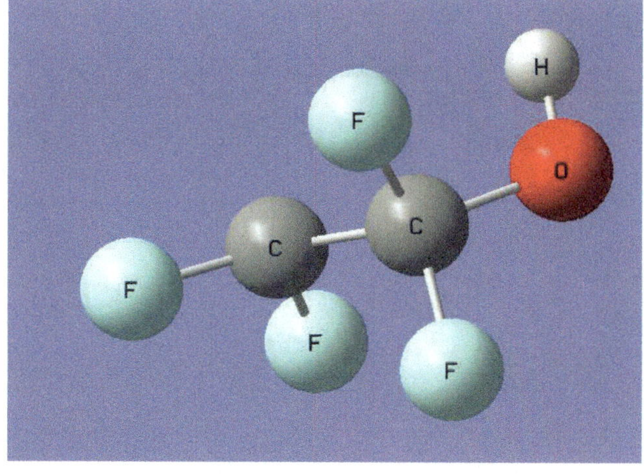

CF$_2$HCF$_2$O•

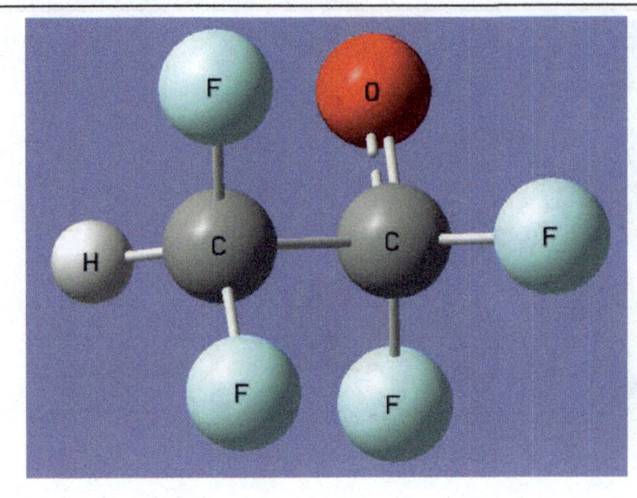

CF$_3$CF$_2$OH

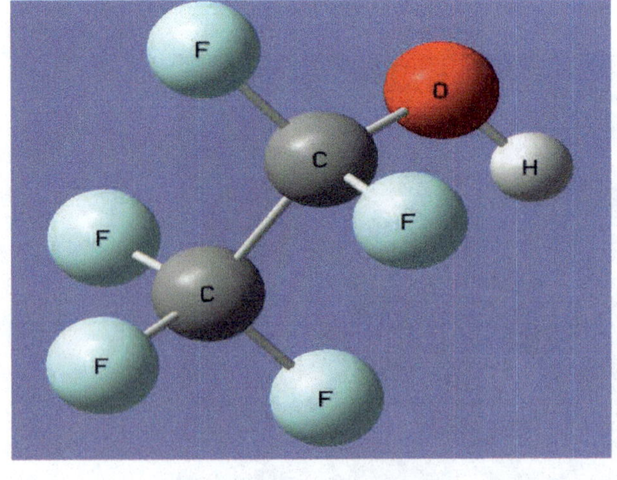

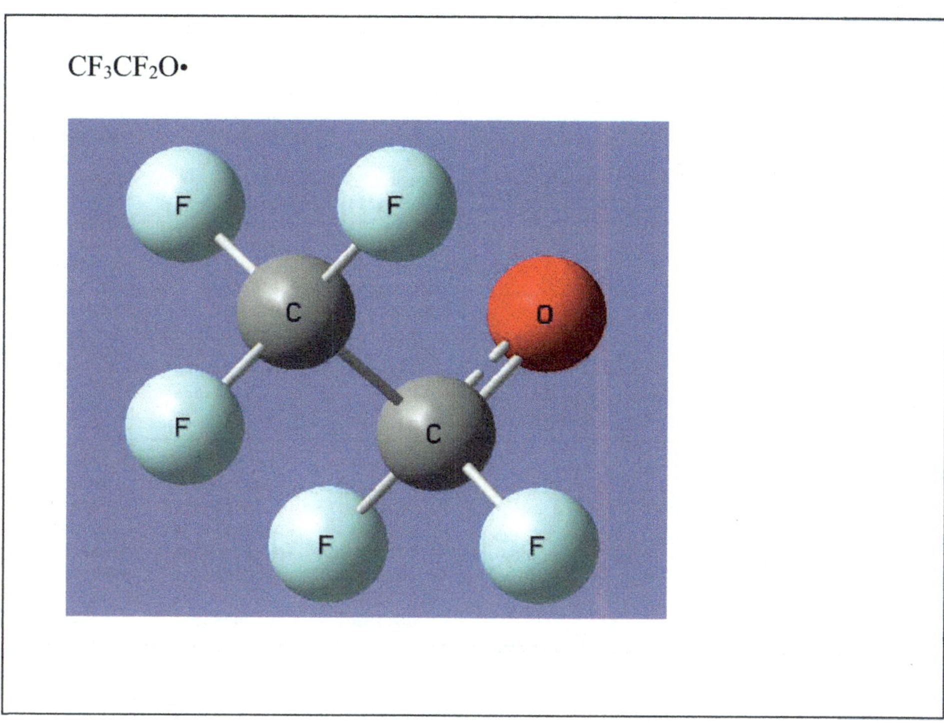

5. Cartesian Coordinates

The Cartesian Coordinates for target tri-, tetra-, and penta- fluorinated ethanol's and their related radicals at the Gaussian m062x/6-31+g (d,p) level of theory are listed in table 5.

Table 5. Cartesian Coordinates in Angstroms for target fluorinated ethanol and their related radical's geometries at the m062x/6-31+g (d,p) level of theory

Molecule	Cartesian Coordinates

| CF3CH2OH | ``` 
 ---------------------------------------------------------------------
 Center     Atomic                    Forces (Hartrees/Bohr)
 Number     Number           X              Y              Z
 ---------------------------------------------------------------------
    1          6         0.000015792    0.000045146    0.000015456
    2          6         0.000056272   -0.000131331   -0.000001091
    3          9        -0.000036015   -0.000005130   -0.000021513
    4          9         0.000035235    0.000054970   -0.000018991
    5          9        -0.000001701    0.000051428    0.000034529
    6          1        -0.000016200    0.000007968    0.000014306
    7          1        -0.000023506   -0.000003074   -0.000011946
    8          8        -0.000022602   -0.000015046    0.000001930
    9          1        -0.000007275   -0.000004931   -0.000012678
 ---------------------------------------------------------------------
``` |
|---|---|
| CF3C•HOH | ```
 ---------------------------------------------------------------------
 Center     Atomic                    Forces (Hartrees/Bohr)
 Number     Number           X              Y              Z
 ---------------------------------------------------------------------
    1          6        -0.000055227   -0.000030630   -0.000017226
    2          6         0.000117503    0.000159657   -0.000153816
    3          9        -0.000061675    0.000035455    0.000071731
    4          9        -0.000053827   -0.000176074    0.000047967
    5          9         0.000080802   -0.000106445    0.000024638
    6          1         0.000001337    0.000046611   -0.000002493
    7          8         0.000004111    0.000030529   -0.000039199
    8          1        -0.000033025    0.000040898    0.000068398
 ---------------------------------------------------------------------
``` |
| CF3CH2O• | ```
                                       -
 ---------------------------------------------------------------------
 Center     Atomic                    Forces (Hartrees/Bohr)
 Number     Number           X              Y              Z
 ---------------------------------------------------------------------
    1          6        -0.000037529    0.000088167    0.000122350
    2          6        -0.000033636   -0.000121335   -0.000217962
    3          9         0.000020892    0.000000631    0.000073365
    4          9         0.000016383   -0.000001502    0.000002646
    5          9        -0.000021622    0.000041297    0.000066627
    6          1        -0.000020139   -0.000044741   -0.000019536
    7          1        -0.000000403   -0.000003543   -0.000012733
    8          8         0.000076054    0.000041026   -0.000014758
 ---------------------------------------------------------------------
``` |
| CHF2CHFOH | ```
 ---------------------------------------------------------------------
 Center     Atomic                    Forces (Hartrees/Bohr)
 Number     Number           X              Y              Z
 ---------------------------------------------------------------------
    1          6         0.000034684   -0.000097837    0.000016859
    2          6         0.000077734   -0.000008144    0.000008869
    3          9        -0.000027798    0.000012351    0.000007573
    4          9        -0.000002203    0.000054383   -0.000025902
    5          1        -0.000018584    0.000015264    0.000009448
    6          9         0.000005269    0.000001678   -0.000006105
    7          9        -0.000005043   -0.000008859    0.000006453
    8          8        -0.000067666    0.000036885   -0.000008025
    9          1         0.000003607   -0.000005722   -0.000009170
 ---------------------------------------------------------------------
``` |

C·F2CHFOH	

```
 -------------------------------------------------------------------
 Center     Atomic                  Forces (Hartrees/Bohr)
 Number     Number          X              Y              Z
 -------------------------------------------------------------------
    1          6        -0.000071214    0.000095917    0.000002631
    2          6        -0.000097297   -0.000427988    0.000070858
    3          9         0.000109064    0.000106013    0.000068648
    4          9         0.000002068    0.000252877   -0.000098474
    5          1        -0.000001211    0.000016670   -0.000018084
    6          9         0.000033735   -0.000023289   -0.000027474
    7          8         0.000054362   -0.000036847    0.000018777
    8          1        -0.000029507    0.000016646   -0.000016882
 -------------------------------------------------------------------
```

CHF2C·FOH	

```
 -------------------------------------------------------------------
 Center     Atomic                  Forces (Hartrees/Bohr)
 Number     Number          X              Y              Z
 -------------------------------------------------------------------
    1          6        -0.000038782    0.000051546    0.000051509
    2          6        -0.000109395   -0.000076573   -0.000071886
    3          9         0.000040508    0.000010628    0.000077586
    4          9         0.000020881    0.000006702   -0.000016053
    5          1         0.000007533   -0.000016674   -0.000041204
    6          9         0.000090694   -0.000038267   -0.000029261
    7          8        -0.000040547    0.000056005    0.000004711
    8          1         0.000029108    0.000006632    0.000024597
 -------------------------------------------------------------------
```

CHF2CHFO·	

```
 -------------------------------------------------------------------
 Center     Atomic                  Forces (Hartrees/Bohr)
 Number     Number          X              Y              Z
 -------------------------------------------------------------------
    1          6         0.000055628   -0.000106091    0.000003381
    2          6         0.000053463   -0.000000718   -0.000007006
    3          9        -0.000043989   -0.000020986    0.000005782
    4          9        -0.000016210   -0.000035868    0.000004447
    5          1        -0.000011020   -0.000013533    0.000007886
    6          1        -0.000010458    0.000024649   -0.000005524
    7          9        -0.000051835    0.000115358   -0.000005383
    8          8         0.000022121    0.000009470   -0.000053144
    9          1         0.000002301    0.000027718    0.000049560
 -------------------------------------------------------------------
```

CH2FCF2OH	

```
 -------------------------------------------------------------------
 Center     Atomic                  Forces (Hartrees/Bohr)
 Number     Number          X              Y              Z
 -------------------------------------------------------------------
    1          6         0.000135651    0.000057453    0.000051033
    2          6        -0.000243304    0.000078716   -0.000013330
    3          9        -0.000116433    0.000005813    0.000097761
    4          1        -0.000077576   -0.000009293    0.000043932
    5          1        -0.000038364   -0.000018670    0.000006345
    6          9         0.000035849    0.000027631   -0.000096988
    7          9         0.000045365    0.000036806   -0.000102497
    8          8         0.000181449   -0.000179561   -0.000003213
    9          1         0.000077364    0.000001106    0.000016957
 -------------------------------------------------------------------
```

C•HFCF2OH

```
 ---------------------------------------------------------------------
 Center     Atomic                     Forces (Hartrees/Bohr)
 Number     Number             X                Y                Z
 ---------------------------------------------------------------------
    1          6         -0.000009430      0.000025594      0.000005992
    2          6          0.000017955     -0.000004162      0.000000569
    3          9          0.000001322      0.000052928     -0.000001501
    4          1         -0.000003779      0.000024041      0.000008036
    5          9         -0.000012190     -0.000046066      0.000011421
    6          9         -0.000025702     -0.000032617      0.000028822
    7          8          0.000010500     -0.000018093     -0.000017353
    8          1          0.000021325     -0.000001626     -0.000035986
 ---------------------------------------------------------------------
```

CH2FCF2O•

```
 ---------------------------------------------------------------------
 Center     Atomic                     Forces (Hartrees/Bohr)
 Number     Number             X                Y                Z
 ---------------------------------------------------------------------
    1          6          0.000157230     -0.000033765      0.000067368
    2          6         -0.000084968     -0.000038414     -0.000127598
    3          9         -0.000060007      0.000028432      0.000052860
    4          1         -0.000002632     -0.000019789     -0.000013222
    5          1         -0.000027022      0.000016735     -0.000013331
    6          9         -0.000031881     -0.000031453      0.000013157
    7          9          0.000080040      0.000071255      0.000005918
    8          8         -0.000030759      0.000006999      0.000014849
 ---------------------------------------------------------------------
```

CF3CFHOH

```
 ---------------------------------------------------------------------
 Center     Atomic                     Forces (Hartrees/Bohr)
 Number     Number             X                Y                Z
 ---------------------------------------------------------------------
    1          6         -0.000139743     -0.000019177     -0.000028107
    2          6         -0.000044583     -0.000023228     -0.000002049
    3          9          0.000017824      0.000030702      0.000057967
    4          9         -0.000069952     -0.000066058     -0.000012198
    5          9          0.000058051     -0.000059520     -0.000018868
    6          9          0.000166545      0.000025560     -0.000007913
    7          1          0.000003967      0.000022634      0.000010499
    8          8         -0.000043206      0.000028605     -0.000066782
    9          1          0.000051097      0.000060482      0.000067450
 ---------------------------------------------------------------------
```

CF3C•FOH

```
 ---------------------------------------------------------------------
 Center     Atomic                     Forces (Hartrees/Bohr)
 Number     Number             X                Y                Z
 ---------------------------------------------------------------------
    1          6          0.000033580      0.000004460      0.000005
    2          6          0.000020098      0.000029526     -0.000031
    3          9          0.000021581     -0.000000752      0.000016
    4          9          0.000016549      0.000006426      0.000007
    5          9         -0.000041946      0.000031215      0.000004
    6          9         -0.000000439     -0.000039185     -0.000010
    7          8         -0.000042830     -0.000023720      0.000011
    8          1         -0.000006593     -0.000007971     -0.000004
 ---------------------------------------------------------------------
```

CF3CFHO•	Center Number	Atomic Number	Forces (Hartrees/Bohr) X	Y	Z
	1	6	0.000536610	0.000642487	-0.00023858
	2	6	-0.000336421	-0.000010459	-0.00005053
	3	9	0.000179754	0.000022776	0.00006963
	4	9	0.000013212	-0.000037718	0.00003827
	5	9	0.000033732	0.000056583	-0.00004308
	6	9	-0.000277185	-0.000172926	0.00005338
	7	1	-0.000020915	-0.000096792	0.00001147
	8	8	-0.000128787	-0.000403952	0.00015944
CF2HCF2OH	Center Number	Atomic Number	Forces (Hartrees/Bohr) X	Y	Z
	1	6	0.000034684	-0.000097837	0.000016859
	2	6	0.000077734	-0.000008144	0.000008869
	3	9	-0.000027798	0.000012351	0.000007573
	4	9	-0.000002203	0.000054383	-0.000025902
	5	1	-0.000018584	0.000015264	0.000009448
	6	9	0.000005269	0.000001678	-0.000006105
	7	9	-0.000005043	-0.000008859	0.000006453
	8	8	-0.000067666	0.000036885	-0.000008025
	9	1	0.000003607	-0.000005722	-0.000009170
C•F2CF2OH	Center Number	Atomic Number	Forces (Hartrees/Bohr) X	Y	Z
	1	6	0.000024554	0.000023392	0.000004415
	2	6	-0.000034477	0.000024861	-0.000019704
	3	9	0.000055726	0.000012426	0.000032150
	4	9	0.000019915	0.000028025	-0.000017047
	5	9	0.000004005	-0.000011188	-0.000003720
	6	9	0.000015833	-0.000043273	0.000042429
	7	8	-0.000068386	-0.000006654	0.000010769
	8	1	-0.000017169	-0.000027587	-0.000049293
CF2HCF2O•	Center Number	Atomic Number	Forces (Hartrees/Bohr) X	Y	Z
	1	6	0.000006128	-0.000032081	0.000064624
	2	6	0.000009899	-0.000032233	-0.000045587
	3	9	-0.000037818	-0.000059393	0.000013603
	4	9	0.000041223	0.000037095	-0.000050906
	5	1	-0.000021281	0.000006425	-0.000018170
	6	9	-0.000013457	0.000031071	0.000014828
	7	9	0.000079466	0.000042020	-0.000014271
	8	8	-0.000064160	0.000007097	0.000035879

CF3CF2OH	Center Number	Atomic Number	Forces (Hartrees/Bohr)		
			X	Y	Z
	1	6	-0.000029279	0.000029059	-0.000014751
	2	6	0.000037295	0.000003602	0.000009039
	3	9	-0.000004027	-0.000031190	-0.000022595
	4	9	-0.000016447	-0.000026160	-0.000006065
	5	9	0.000008918	-0.000023108	0.000015387
	6	9	-0.000006450	0.000029260	-0.000000832
	7	9	0.000001454	0.000001809	-0.000025482
	8	8	-0.000002384	0.000003392	0.000022090
	9	1	0.000010920	0.000013336	0.000023209

CF3CF2O•	Center Number	Atomic Number	Forces (Hartrees/Bohr)		
			X	Y	Z
	1	6	0.000090683	0.000175823	0.000176247
	2	6	-0.000004772	-0.000005489	-0.000074419
	3	9	-0.000035079	0.000031055	-0.000015124
	4	9	-0.000026035	-0.000069252	0.000016300
	5	9	0.000038428	-0.000012873	-0.000000972
	6	9	-0.000080090	-0.000021765	-0.000033887
	7	9	0.000031147	-0.000065731	-0.000034377
	8	8	-0.000014283	-0.000031768	-0.000033768

6. Conclusion

The ab initio and Global-hybrid meta-GGA density function method, Gaussian M06-2x/6-31+g (d,p), used to calculate thermodynamic properties of 18 tri-, tetra-, and penta- fluorinated ethanol's and their related radicals are presented. Calculated standard enthalpy of formation at the Gaussian M06-2x/6-31+g (d,p) using multiple work reactions are also presented; work reactions were employed for cancellation of calculation errors, table2.

The calculated Thermochemical properties: the O—H, secondary methyl C—H, and ethyl C-H Bond Dissociation Energies (BDEs), and Standard Enthalpy of formation (298K) values for tri-, tetra-, and penta-fluorinated Ethanol's and their related Radicals: CH3CH2-xFxOH, CH3-xFxCH2-xFxOH, and CH2-xFxCH2OH are tabulated, table 3.

The C-H bond dissociation energies ranged from 98.6 to 104.9 Kcal mol^{-1} on the secondary ethyl carbons and from 103.3 to 106.7 Kcal mol^{-1} on the primary methyl carbons. The O-H bond energies range from 110.7 to

118.9 Kcal mol^{-1}. The calculated O-H bond energies increased by intruding more fluorine atoms into either methyl or ethyl carbons.

Chapter 10
The Effect of Lack of Oxygen and Excess Carbon Dioxide Buildup on Joint Diseases and The Natural Treatments for Some Joint Diseases. A Review

1. Overview

The leading cause of disability worldwide is Arthritis. Arthritis can be either an autoimmune joint disease or osteoarthritis. According to the Center for Disease Control (CDC), by 2040, 80 million U.S. adults will have some form of arthritis. This chapter presents the effect of the obstructed ability to breathe by a physical block limit or by prolonged hypoventilation, or by a pulmonary condition on hypoxia, hypercapnia, and blood chemical PH imbalance. Mechanism of alternating blood PH values, methods of identifying blood chemical PH imbalance, methods of measuring P_{CO2}, P_{O2}, and blood PH, types of joint diseases caused due to blood chemical PH imbalance, current medicine prescribed in their treatments, and the natural treatments of such joint diseases are also discussed.

2. Introduction and Background

The earth's atmosphere contains 21 percent oxygen. A closed space has safe oxygen levels if readings are between 20.8-21 %, while a space with readings of less than 19.5 % are oxygen deficient according to OSHA guidelines. Home air conditioners: split AC, window AC, and portable AC can't ventilate your room: they don't have the capability to bring the outside air; they only circulate indoor air and cool it. Complex HVAC systems used inside some hotels, office buildings, big malls, and airports have the feature of ventilating the indoor area and maintaining freshness, humidity, and oxygen level. Some window ACs in the US can bring outside air and maintain oxygen levels. Most window ACs around the globe do not have this feature. Oxygen room level of 19.5-23.5% is considered safe, and oxygen levels between 14-16% and below 19.5% are considered hazardous.

If, while breathing out, not enough carbon dioxide is expelled from the lungs, the increased carbon dioxide in the blood reduces the blood's PH and makes the blood acidic, causing raspatory acidosis. When your ability to breathe is blocked by physical block limits, another condition or a disease, such as raspatory acidosis, is caused. Raspatory acidosis can be either acute, chronic, or acute and chronic. The sudden arrival of CO2 in the lungs is called Acute Respiratory Acidosis. The kidney's response to acute respiratory acidosis is so quick that it can happen within minutes. The causes may include cerebrovascular accidents like strokes, a group of diseases that interfere with gene's ability to make muscle and causes steadily lose of muscles causing Muscular dystrophy, voluntary muscles may become weak, or loss of control of them causing Myasthenia gravis, heart attack, a very rare neurological disorder where the immune system attacks itself and can cause problems from trouble eating to full body paralysis causing Guillain-Barre Syndrome, and/or block airways.

Chronic raspatory acidosis is more serious, and it happens at a slower rate and at a lesser degree than acute raspatory acidosis. The lower oxygen rate for the tissues to be fully supplied in the blood is known as hypoxemia. The conditions may include pulmonary fibrosis or diseases that happens in the lung tissue/ nerve or muscular diseases, sleep apnea, obesity, thoracic skeletal defects that cause pecs/rib cage/or sternum to be shaped in a way that limits lung functioning or breathing, Asthma, and a group of airflow and breathing diseases including diseases like bronchitis and emphysema called Chronic obstructive pulmonary disease (COPD). Common treatments for raspatory acidosis are oxygen tubes, different medications or other treatments to stop smoking, naloxone (for opioid overdose), anti-inflammatory medications to ease any constrictive swelling, and breathing machines like a CPAP or BiPAP. To prevent getting raspatory acidolysis in general: lose weight if overweight, quit smoking, and don't drink alcohol while taking opioids and strong pain medications.

A two-year study showed that there is an adverse physical effect on medical staff when wearing an N 95 mask. Wearing an N 95 mask resulted in hypercapnia (excessive CO2 in the blood caused by inadequate respiration) and hypoxemia (lack of oxygen in the blood) which reduced the ability to make a correct decision and the working efficiency. Dizziness, shortness of breath, and headache were experienced by the medical staff wearing N95 masks.

If prolonged hypoventilation accompanies raspatory acidosis, the condition becomes more severe, and it can cause the patient to have additional symptoms: myoclonus, seizures, and altered mental status. Hypercapnia (excessive amounts of carbon dioxide in the blood) can be caused by respiratory acidosis leading to cerebral vasodilation. Severe respiratory acidosis may cause papilledema and increased intracranial pressure increasing the risk of death and herniation. Chronic respiratory acidosis may cause impaired coordination, polycythemia, memory loss, heart failure, and pulmonary hypertension.

Dyspnea (difficult respiration) is commonly caused by chronic obstructive pulmonary disease (COPD), interstitial lung disease, heart failure, asthma, psychogenic problems that are usually linked to anxiety, and pneumonia. Chest X-rays and computed tomography (CT) images are used by doctors for its diagnosis. Spirometry tests can be used to measure the airflow and the patient's lung capacity and to pinpoint the extent and the type of an individual's breathing problems. Other tests can be used to directly measure the blood capacity to carry oxygen and the level of oxygen in the blood. Treatment of dyspnea is dependent on its cause. If it's caused from having bacterial pneumonia, antibiotics are prescribed. If it's caused from having asthma, bronchodilators and steroids are prescribed, and other medications are also effective (opiates, anti-anxiety drugs, and non-steroidal anti-inflammatory drugs (NSAIDs)). Special breathing techniques, such as breathing muscle strengthening exercises and pursed-lip breathing, can be used if the cause of dyspnea is caused by COPD. Supplemental oxygen may be prescribed if tests indicate low levels of oxygen in the blood.

A less well-known side effect of using nonsteroidal anti-inflammatory drugs (NSAIDs drugs) is the degradation of joint cartilage. Nonsteroidal anti-inflammatory drugs include; Ibuprofen, used in some drugs such as Mortin, Nuprin, and Advil; Piroxicam, used in drugs such as Feldene; diclofenac, used in drugs such as Voltaren; fenoprofen, used in drugs such as Nalfon; indomethacin used in drugs such as Indocin; naproxen used in drugs such as Naprosyn; tolmetin used in drugs such as Tolectin; and sulindac used in drugs such as Clinoril. Other side effects of using nonsteroidal anti-inflammatory drugs are ulcer formation, dizziness, headaches, and gastrointestinal upset. Clinical studies have shown that nonsteroidal anti-inflammatory drug usage has caused an acceleration of osteoarthritis and increased joint destruction.

Common joint diseases may include bursitis, osteoarthritis, gout, rheumatoid arthritis, spondyloarthritis, lupus, and juvenile idiopathic arthritis. Bursitis is caused by the inflammation of the small, fluid-filled sacs called bursae around the joints, tendons, muscles, and bones. Overuse or sudden injury of joints such as the elbow, hip, and shoulder can lead to flare-ups. Bursitis can sometimes result from bacterial infections. Osteoarthritis is the wear-and-tear form that increases with age. Adults in their 50s and older are more likely to develop this chronic and progressive disease. Women are more vulnerable to osteoarthritis than men. It's stiffness and pain with movement caused by breaks down of cartilages that cushion the joints. The flexibility decreases, and walking becomes more difficult, especially with knee and hip arthritis. The type of arthritis that affects the joint connecting the big toe to the rest of the foot is called gout. A waste product in the blood, uric acid, would exist in excess and forms crystals in the joints. Flare-ups caused by gout are extremely painful and it would commonly strike at night. Men are more vulnerable to gout disease, and women become more vulnerable to this disease after menopause.

The autoimmune condition that affects the lining of the joints is called Rheumatoid arthritis. Immune system cells accumulate in large numbers in the joints. The interaction between joint cells and immune cells causes increased inflammation leading to damage and destruction of the bone and cartilage. Spondyloarthritis consists of certain other rheumatoid diseases, including axial spondylitis, enteropathic arthritis, and psoriatic arthritis. The inflammation in the spine that can eventually lead to spinal fusion or ankylosing spondylitis is called axial spondylitis. The complication of inflammatory bowel diseases, like ulcerative colitis, is called enteropathic arthritis. Psoriatic arthritis is associated with the skin condition psoriasis, and it tends to affect the joints of the hands and feet. The autoimmune condition affects various parts of the body, including internal organs, skin, brain, blood, bones, and joints are called lupus. Lupus can cause an inflammation that triggers arthritis, particularly in the knees, feet, hands, elbows, and shoulders. The most common chronic joint condition in kids is Juvenile idiopathic arthritis. The child's immune system attacks the body's own healthy tissue, and it's an autoimmune condition. The cause is unknown, and it may alter children's normal growth. This inflammation may affect the internal organs, eyes, muscles, joints, and ligaments.

3. Review and Discussion

The amount of oxygen circulating in the blood is the blood oxygen level. The normal reading using an oximeter is between 95 to 100. Forms of carbon dioxide carried in the blood are carbaminohemoglobin (CO_2 bound to hemoglobin) and chemically modified bicarbonate (HCO_3^-). The solubility of CO_2 in the blood is 0.07 mL CO_2/100 mL blood/mm Hg which would be almost 5% of the total CO_2 content of the blood. The solubility of oxygen in the blood is 0.003 mL O_2/100 mL blood/mm Hg, which would be almost 2% of the total O_2 content of the blood. The solubility of Carbon dioxide is 20 times more soluble in blood than oxygen. According to Henry's law, the solubility of a gas is directly proportional to the partial pressure of that gas above the liquid. Increasing the pressure and decreasing the temperature would lead to an increase in the solubility of gaseous over a liquid. 16 Carbon dioxide gas has a lower ability to diffuse and exit the lungs compared to oxygen gas according to Graham's law 95as it's denser than oxygen gas, the density of oxygen gas is 1.43 g/L compared to the density of carbon dioxide gas 1.81 g/L.

The body normally maintains CO_2 in a range from 38 to 42 mm Hg by balancing its elimination and production. Ventilation is primarily initiated by the blood PH. The PH of the blood is mainly regulated by the amount of CO_2 in the blood. The body produces more CO_2 than it can eliminate in case of hypoventilation, causing retention of CO_2. The increased CO_2 is what leads to an increase in blood acidity due to an increase of hydrogen ion concentration, and a slight increase in bicarbonate concentration, and the equilibrium would shift toward forming more hydrogen ions according to the following reaction:

$$CO_2 + H_2O \rightarrow H_2CO_3 \rightarrow HCO_3^- + H^+$$

A buffer system is created from the presence of the flowing molecules: HCO_3, CO_2, and $H_2CO_3^-$ in equilibrium.

In the presence of excess hydroxide ions (OH-), carbonic acid (H2CO3) would buffer a high PH, and in the presence of excess hydrogen ions (H+), carbonate anion (HCO3-) would buffer a low PH which is the main mechanism behind raspatory acidolysis blood PH drop. In raspatory acidolysis, the slight increase in bicarbonate act as a buffer for the increase in H+ ions, which helps minimize the drop in PH blood value. Increased hydrogen ions (H+) would lead to a slight decrease in the buffered blood PH; blood PH would be below 7.35.

To evaluate patients with suspected respiratory acidosis, serum bicarbonate level and an arterial blood gas (ABG) are necessary. An elevated bicarbonate level of HCO3- (>30 mmHg), an elevated PCO2 (>45 mmHg), and decreased pH (<7.35) would show on an ABG test in case of raspatory acidolysis. Respiratory acidosis can be either chronic or acute based on the relative increase in HCO3- with respect to PCO2. An HCO3- will have increased by one mEq/L for every ten mmHg increase in PCO2 over a few minutes in case of acute respiratory acidosis. An HCO3- will have increased by four mEq/L for every ten mmHg increase in PCO2 over a time course of days in case of chronic respiratory acidosis. A mixed respiratory-metabolic disorder may be present if it doesn't show either pattern of acute or chronic respiratory acidosis. A drug screen may also be warranted in a patient who shows unexplained respiratory acidosis.

The cause of respiratory acidosis must be treated once the diagnosis has been made. The rapid alkalization of the cerebrospinal fluid (CSF) may lead to seizures; therefore, the hypercapnia should be corrected gradually. To help improve ventilation, pharmacologic therapy may be used. Beta-agonists, anticholinergic drugs, and methylxanthines (Bronchodilators) may be used in treating patients with obstructive airway diseases. In case of patients who overdose on opioid use, Naloxone can be prescribed.

Acidolysis can be either raspatory or metabolic in origin, depending on the measured pCO2. If pCO2 is greater than 40 to 45, it's known to be due to decreased ventilation, and it's called respiratory acidosis. If the pCO2 is less than 40 since it is not the cause of the primary acid-base disturbance, it's called metabolic acidosis, and it's confirmed by a measured decrease in bicarbonate (normal range 21 to 28 mEq/L).

In alkalosis, the fluids of the body are alkaline; the blood PH is high. When the blood has too little acid making it basic, the condition is called alkalosis; blood PH would be higher than the normal PH value of 7.45. There might be no noticeable symptoms for mild and chronic alkalosis. If there is a rapid PH increase, symptoms may include confusion, nausea or vomiting, muscle twitching or spasms, numbness of the hands and feet, and/ or dizziness or lightheadedness.

Alkalosis can be either raspatory or metabolic, depending on pCO2. If pCO2 is greater than 45 mmHg, it's called metabolic alkalosis, and the measured bicarbonate is greater than 29mM. If pCO2 is less than 32 mmHg, it's called raspatory alkalosis, and the measured bicarbonate is less than 22mM.

Figure 1. Types of Blood Alkalosis and Acidosis.

	pH	pCO$_2$	Total HCO$_3^-$
Metabolic acidosis	↓	N, then ↓ ↓	
Respiratory acidosis	↓	↑	N, then ↑
Metabolic alkalosis	↑	N, then↑	↑
Respiratory alkalosis	↑	↓	N, then ↓

Reference values (arterial): pH: 7.35–7.45; pCO$_2$: male: 35–48 mm Hg, female: 32–45 mm Hg; total venous bicarbonate: 22–29 mM. N denotes normal; ↑ denotes a rising or increased value; and ↓ denotes a falling or decreased value.

A number of diseases, including rheumatoid arthritis (RA), are caused by alterations in tissue oxygen pressure. In a condition known as hypoxia, low partial pressure of oxygen is involved in angiogenesis, apoptosis, cartilage degradation, inflammation, oxidative damage, and energy metabolism. Synovial hypoxia can be linked to pathogenic processes through indirect and direct effects on angiogenesis, oxidative damage, inflammation, cartilage damage, and bone resorption. Studies show that hypoxia and other promoters lead to inflammation in Rheumatoid Arthritis. The metabolic environment in the synovium is modified by hypoxia, and an autoimmune response is initiated by the presentation of the upregulated antigenic enzymes in the context of cellular stress. Hypoxia induces an anaerobic glycolytic phenotype in the synovium. Several of the enzymes induced by this metabolic shift may become antigenic once anaerobic glycolysis is established in the synovium.

The abnormal biomechanics, attendant tissue-derived, and cell-derived factors cause Osteoarthritis (OA). The progression of Osteoarthritis is related to reactive oxygen species (ROS) and oxidative stress. It's a multifactorial and polygenic joint disease. Reactive oxygen species target the complex oxidative stress signaling pathways as it regulates chondrocyte senescence and apoptosis, extracellular matrix synthesis and degradation, along with synovial inflammation and dysfunction of the subchondral bone and intracellular signaling processes. Osteoarthritis progresses from silent cartilage destruction to painful presentation. Free radicals mediate and amplify the sequence of joint degeneration in all tissues affected due to their chemical properties. Free radicals are the crucial factor involved in the inflammatory transformation of Osteoarthritis joints and all joint tissues disease development. Both Osteoarthritis and Rheumatoid arthritis are caused by reduced oxygen levels from increased consumption by inflammatory cells such as synoviocytes and the oxygen-reduced delivery to synovial fluid.

Gout is a result of the presence of uric acid crystals building up in the joints. The production of uric acid in the joints is increased by the presence of excess carbon dioxide and the lack of presence of oxygen in the lungs. While sleeping, it produces less cortisone, an inflammation suppressant; the reduction of cortisone level might be contributing to gout disease. A person may have as much as a 50% chance of getting gout disease if they suffer from sleep apnea. Dehydration and loss of water in the body can contribute to the increase in the concentration of uric acid in joint fluids, enhancing the formation of uric acid crystals in the joints and causing gout attacks. Gout is a type of arthritis that is caused by a uric acid concentration increase in the blood; it may cause debilitation due to uric acid deposits around tendons and joints. It's the most controllable metabolic disease. Gout is classified into primary and secondary. Primary gout causes are unknown. There are known genetic defects causing elevated uric acid. The increase of uric acid in primary gout can be due to the reduced ability to excrete uric acid found in a smaller group of patients (30%), increased formation of serum uric acid found in most patients, or both, which is found in a minority of patients.

Rheumatoid arthritis can't be cured by drugs, but it can be treated as it can/will come back. Treatment of rheumatoid arthritis includes using medications to slow the progression of the disease. Drugs including sulfasalazine (Azulfidine), methotrexate (Trexall), and other biologic drugs such as etanercept (Enbrel) and adalimumab (Humira) may be prescribed. Biologic drugs reduce inflammation by targeting the immune system. Short-term treatment may include low-dose steroids. Rheumatoid Arthritis is a multifactorial disease as both environmental and genetic factors contribute to the disease. Medical therapy is limited in most RA cases; it fails to address the causes of the disease. As in Osteoarthritis, the use of NSAIDs drugs, including aspirin, is accompanied by the acceleration of factors that promote the disease process. Examples of drugs currently in use are hydroxychloroquine, penicillamine, methotrexate, gold therapy, azathioprine, and cyclophosphamide. A diet rich in vegetables, fiber, and whole foods and low in meat, sugar, saturated fat, and refined carbohydrates (Western Diet) prevents the development of Rheumatoid Arthritis disease. Dietary therapy is to follow a vegetarian diet, eliminate food allergies, increase the intake of antioxidants, and alter the intake of fats and oils. Dietary therapy shows tremendous promise in the treatment of Rheumatoid Arthritis. Incomplete digestion may be a major factor in Rheumatoid Arthritis. Gamma-Linolenic acid (GLA) acts as a precursor to an anti-inflammatory prostaglandins' series 1. Studies show that some patients have responded to GLA treatment while others didn't. Fish oil supplementation containing Omega-3 Fatty acids shows a better and more positive response than GLA in the treatment of Rheumatoid Arthritis. Due to the neutralization of inflammation and support of collagen structure, dietary antioxidants such as Flavonoids are used in the treatment of Rheumatoid Arthritis. Patients with Rheumatoid Arthritis have low levels of selenium. Selenium combined with Vitamin E had a positive effect in the treatment of Rheumatoid Arthritis. Zinc levels are commonly low in Rheumatoid Arthritis patients; treatments with zinc supplements in the form of sulfates showed a slight therapeutic effect. Patients with Rheumatoid Arthritis are deficient in manganese-containing Superoxide dismutase. The injectable form of the enzyme (antioxidant enzyme Superoxide dismutase (Manganese SOD) available in Europe is effective in the treatment of Rheumatoid Arthritis. Oral administration of SOD showed no effect. Patients with Rheumatoid Arthritis are also deficient in vitamin C. Supplements with vitamin C give some anti-inflammatory action. Pantothenic acid in blood has been shown to be lower in Rheumatoid Arthritis patients. Patients who received 2g of calcium Pantothenate daily showed improvement. Arthritis patients showed a lower sulfur content in the fingernails. Using injectable sulfur alleviated pain and swelling. A high dose of Niacinamide (900-4000mg) has shown good results in the treatment of both Osteoarthritis and Rheumatoid Arthritis. The administration of 500 mg of Pantothenic acid has shown no effect on the treatment of Rheumatoid Arthritis.

A vegetarian diet following short-term fasting showed a reduction of Rheumatoid Arthritis disease activity.

Treatment of Osteoarthritis includes prescribing medications called bisphosphonates: risedronate (Actonel) and alendronate (Fosamax). In most Osteoarthritis cases, drug treatment has shown to be ineffective, and if the failure of nonsurgical treatment is consistent in at least three to six months, surgical replacement of large joints: knee replacement, or hip replacement, is needed.

The therapeutic goal in the natural treatment of osteoarthritis is to enhance and repair the collagen matrix and the regeneration of the connective tissues. It's recommended to lose excess weight, causing increased stress on joints, use a healthy diet rich in complex carbohydrates and fiber, and minimize and eliminate the consumption of nightshade vegetables. Nightshade vegetables, including potatoes, tomatoes, peppers, tobacco, and eggplants, are known to contain alkaloids that promote the inflammatory degradation of the joints and inhibit normal collagen repair. The high intake of antioxidants is shown to inhibit the progression of the disease and reduce the risk of cartilage loss. As some people age, they lose their ability to manufacture sufficient levels of glucosamine. Glucosamine, in the form of glucosamine sulfate drug, is used to incorporate sulfur into the cartridges and stimulate the manufacture of glucosamineglycans.

Chondroitin sulfate is a drug that is composed of repeating units of derivatives of glucosamine sulfate attached to sugar molecules and is known to be a less effective drug than that of glucosamine sulfate due to its low absorption 0-13% compared to the solubility of glucosamine sulfate which is 90-98%. A high dose of Niacinamide (900-4000mg) has shown good results in the treatment of both Osteoarthritis and Rheumatoid Arthritis. Superoxide Dismutase (SOD) injections showed a significant effect in the treatment of osteoarthritis. Vitamin E has the ability to stimulate the formation of new cartilage components and inhibit the breakdown of cartilage, and the administration of 600 IU showed significant benefits. Vitamin C is like vitamin E and protects and enhances cartilage formation. The administration of a small amount (12.5mg) of Pantothenic acid is effective in relieving symptoms of Osteoarthritis. Joint degradation is accelerated by the deficiency of one of the vitamins: A, B6, Zinc, Copper, and Boron, and supplementation at the appropriate level may promote cartilage synthesis and repair. Niacinamide in high dosage of 900 to 4000mg per day showed a promising result for the treatment of Osteoarthritis.

Treatments of gout Joint disease include prescribed medications, such as allopurinol and febuxostat. Treatment of sleep apnea that would include continuous positive airway pressure (CPAP) machine or another treatment device to increase oxygen intake while sleeping is used to increase oxygen level and lower uric acid production, and reduce the risk of a gout attacks. Decreasing the concentration of uric acid by drinking fluids would increase blood volume and lowers the risk of gout attacks. Other lifestyle changes that may lower the risk of a gout attack include getting regular exercise, eating a plant-based diet that is low in purines and whole foods, and losing excess weight. 93,94. The standard medical treatment for the disease is the administration of Colchicine, indomethacin, naproxen, fenoprofen, or phenylbutazone. The dietary treatment involves fluid intake, low fat, low purine, and low protein intake, consumption of complex carbohydrates, elimination of alcohol intake, and achieving the ideal body weight. The natural treatment of gout disease includes the consumption of nutritional supplements: eicosapentaenioc acid. Vitamin E, folic acid, amino acids such as alanine, aspartic acid, glutamic acid, glycine, and niacin, and vitamin C. Eicosapentaenioc acid, omega-3 oils are found useful in the treatment of gout joint disease. Vitamin E acts as an antioxidant and inhibits the formation of leukotrienes. Folic acid is known to inhibit the production of uric acid by inhibiting the enzyme xanthine oxidase. Alanine, aspartic acid, glutamic acid, and glycine these amino acids are shown to lower serum. Uric acid level by increasing uric acid excretion. Niacin and Vitamin C, High doses of vitamin C and Niacin are used in the treatment of gout; niacin competes with uric acid in excretion, and vitamin C increases the formation of uric acid in small groups of people.

4. Conclusion

If the ability to breathe is obstructed by a pulmonary condition or by, physical block limits, or prolonged hypoventilation, raspatory acidosis is caused. Three main types of joint diseases are directly related to hypoxia (low levels of oxygen in your body tissues), hypercapnia (excessive amounts of carbon dioxide in the blood), and blood PH imbalance: rheumatoid arthritis, osteoarthritis, and gout joint diseases.

Both osteoarthritis and rheumatoid arthritis are caused by reduced oxygen levels from increased consumption by inflammatory cells such as synoviocytes and the oxygen-reduced delivery to synovial fluid. Rheumatoid arthritis is directly caused by alterations in tissue oxygen pressure. Osteoarthritis is directly related to aging, as the ability to synthesize and restore cartilage structure decreases, and the progression of the disease was found to be directly related to the presence of reactive oxygen species and oxidative stress. Gout joint disease is directly caused by the presence of uric acid crystals building up in the joints, and the crystal buildup is increased by the presence of excess carbon dioxide.

The natural treatment of both rheumatoid arthritis and osteoarthritis includes minimizing the consumption of nightshade vegetables and eliminating the use of nonsteroidal anti-inflammatory drugs (NSAIDs drugs) as they cause the degradation of joint cartilages. A high dose of Niacinamide (900-4000mg) has shown good results in the treatment of both Osteoarthritis and Rheumatoid Arthritis. Vitamin C is, like vitamin E supplements, has shown good results in the treatment of osteoarthritis and rheumatoid arthritis, and gout joint diseases. It's recommended to increase oxygen intake while sleeping and increase blood volume by drinking fluids to lower the risk of a gout attack. It's recommended to lose excess weight and maintain a healthy body weight in the treatment of both Osteoarthritis and Gout joint diseases.

The dietary treatment of Rheumatoid Arthritis includes following a vegetarian diet, eliminating food allergies, increasing the intake of antioxidants, and altering the intake of fats and oils, dietary treatment of Osteoarthritis includes the use of a healthy diet rich in complex carbohydrates and fiber, to minimize and eliminate the consumption of nightshade vegetables, and to intake antioxidants, and dietary treatment of Gout joint disease includes fluid intake, low fat, low purine, and low protein intake, consumption of complex carbohydrates, and elimination of alcohol intake.

The natural treatment of gout joint disease includes the administration of nutritional supplements: eicosapentaenioc acid. Vitamin E, folic acid, amino acids such as alanine, aspartic acid, glutamic acid, glycine, and niacin, and vitamin C, the natural treatment of rheumatoid arthritis disease includes the administration of some nutritional supplements: Omega-3 Fatty acid, Selenium combined with vitamin E, vitamin C, Niacinamide, and Pantothenic acid, and the natural treatment of osteoarthritis disease include the administration of nutritional supplements: glucosamine sulfate, Niacinamide, Vitamin E, Vitamin C, small amounts of Pantothenic acid, and Niacinamide.

References

Chapter 1. Chemical Formulations for Mouthwashes

1. https://www.westenddental.com/blog/is-mouthwash-necessary/
2. Johansson I, Somasundaran P, Handbook for Cleaning/ Decontamination of Surface, Elsavier, Kidlington, Oxford OX5, 2007. Pp 382-383.
3. Manton David J., Hayes-Cameron L., Handbook of Pediatric Dentistry 4th edition, 2013, Dental Caries, Mosby Elsevier, Canberra Australia. pp-49-78.
4. https://www.ada.org/resources/research/science-and-research-institute/oral-health-topics/mouthrinse-mouthwash
5. Torres CR, Perote LC, Gutierrez NC, Pucci CR, Borges AB. Efficacy of mouth rinses and toothpaste on tooth whitening. Oper Dent 2013;38(1):57-62.
6. Marinho VC, Higgins JP, Logan S, Sheiham A. Topical fluoride (toothpastes, mouthrinses, gels or varnishes) for preventing dental caries in children and adolescents. Cochrane Database Syst Rev 2003(4):CD002782.
7. Hasson H, Ismail AI, Neiva G. Home-based chemically-induced whitening of teeth in adults. Cochrane Database Syst Rev 2006(4):CD006202.
8. Sharma N, Charles CH, Lynch MC, et al. Adjunctive benefit of an essential oil-containing mouthrinse in reducing plaque and gingivitis in patients who brush and floss regularly: a six-month study. J Am Dent Assoc 2004;135(4):496-504
9. Blom T, Slot DE, Quirynen M, Van der Weijden GA. The effect of mouthrinses on oral malodor: a systematic review. Int J Dent Hyg 2012;10(3):209-22.
10. Araujo MW, Charles CA, Weinstein RB, et al. Meta-analysis of the effect of an essential oil-containing mouthrinse on gingivitis and plaque. J Am Dent Assoc 2015;146(8):610-22.
11. Sharma N, Charles CH, Lynch MC, et al. Adjunctive benefit of an essential oil-containing mouthrinse in reducing plaque and gingivitis in

patients who brush and floss regularly: a six-month study. J Am Dent Assoc 2004;135(4):496-504

12. Fedorowicz Z, Aljufairi H, Nasser M, Outhouse TL, Pedrazzi V. Mouthrinses for the treatment of halitosis. Cochrane Database Syst Rev 2008(4):CD006701.

13. Leary, Kecia S., Nowak, Arthur J., Pediatric dentistry: infancy through adolescence, Prevention of Dental Disease, P455-460; 2019.

14. Mariotti AJ, Burrell, K.H. Mouthrinses and Dentifrices. 5th ed. Chicago: American Dental Association and Physician's Desk Reference, Inc.; 2009.

15. Mariotti AJ, Burrell, K.H. Mouthrinses and Dentifrices. 5th ed. Chicago: American Dental Association and Physician's Desk Reference, Inc.; 2009.

16. Hasson H, Ismail AI, Neiva G. Home-based chemically-induced whitening of teeth in adults. Cochrane Database Syst Rev 2006(4):CD006202.

17. Kerr AR, Corby PM, Kalliontzi K, McGuire JA, Charles CA. Comparison of two mouthrinses in relation to salivary flow and perceived dryness. Oral Surg Oral Med Oral Pathol Oral Radiol 2015;119(1):59-64.

18. https://www.animated-teeth.com/bad_breath/t5_halitosis_cures.htm

19. https://www.webmd.com/diet/distilled-water-overview#1

20. https://www.animated-teeth.com/bad_breath/t5_halitosis_cures.htm

21. https://pubchem.ncbi.nlm.nih.gov/compound/Castor-oil

22. Andrade IM, Andrade KM, Pisani MX, et al. Trial of an experimental castor oil solution for cleaning dentures. *Braz Dent J.* 2014;25(1):43-47.

23. Badaró MM, Salles MM, Leite VMF, et al. Clinical trial for evaluation of Ricinus communis and sodium hypochlorite as denture cleanser. *J Appl Oral Sci.* 2017;25(3):324-334.

24. Salles MM, Badaró MM, Arruda CN, et al. Antimicrobial activity of complete denture cleanser solutions based on sodium hypochlorite and Ricinus communis – a randomized clinical study. *J Appl Oral Sci.* 2015;23(6):637-642.

25. Vieira C, Evangelista S, Cirillo R, et al. Effect of ricinoleic acid in acute and subchronic experimental models of inflammation. *Mediators Inflamm.* 2000;9(5):223-228.

26. 26.*Gosselin, R.E., H.C. Hodge, R.P. Smith, and M.N. Gleason. Clinical T oxicology of Commercial Products. 4th ed. Baltimore: Williams and Wilk ins, 1976., p. II-152*

27. https://en.wikipedia.org/wiki/Peppermint_extract

28. https://en.wikipedia.org/wiki/Sodium_benzoate

29. Penniston K.L., Nakada S.Y., Holmes R.P., Assimos D.G., Quantitative Assessment of Citric Acid in Lemon Juice, Lime Juice, and Commercially-Available Fruit Juice Products, National Library of Medicine, Journal of Endourology, 2008;22(3):567-70.

30. https://pubchem.ncbi.nlm.nih.gov/compound/Citric-acid

31. https://www.glowdental.co.uk/how-citric-acid-affects-your-teeth/

32. https://en.wikipedia.org/wiki/Tetrasodium_EDTA

33. https://www.lorealparisusa.com/ingredient-library/disodium-edta

34. https://www.hsph.harvard.edu/nutritionsource/food-features/coconut-oil/

35. https://www.ncbi.nlm.nih.gov/pmc/articles/PMC4625327/

36. https://pubchem.ncbi.nlm.nih.gov/compound/Water

37. https://en.wikipedia.org/wiki/Sodium_chlorite

38. https://en.wikipedia.org/wiki/Castor_oil

39. https://pubchem.ncbi.nlm.nih.gov/compound/Peppermint-oil

40. https://pubchem.ncbi.nlm.nih.gov/compound/Sodium-benzoate

41. https://pubchem.ncbi.nlm.nih.gov/compound/Castor-oil

42. https://pubchem.ncbi.nlm.nih.gov/compound/Lemongrass-oil

43. Charles C.H, Cronin M. J., Dembling W. Z., Petrone D. M., McGuire J. A., Anticalculus efficacy of an antiseptic mouthrinse containing zinc chloride, Journal of American Dental Association, 2001;132(1):94-8, doi: 10.14219/jada.archive.2001.0033.

44. https://usatoday30.usatoday.com/news/health/2008-05-13-bad-breath_N.htm

45. http://dentalingredients.spectrumchemical.com/products/oral-care-flavorants

Chapter 2. Chemical Formulations for Acrylic Matt and Acrylic Gloss Paints

1. http://pscf.com.sa/

2. The Essential Chemical Industry, 2013, Paints. Retrieved February 9, 2022, from www.essentialchemicalindustry.org

3. Dunn Edwards Paints, 2013, What is Paints Made of? Retrieved February 10, 2022, from www.dunnedwards.com.

4. Wikipedia, January 2022, Surfactants in Paint. Retrieved February 9, 2022, from https://en.wikipedia.org/

5. Ebbing and Gammon, 2015, General Chemistry 11th edition, Boston MA, Cengage.

6. https://coatings.specialchem.com/product/a-ashland-specialtychemical-natrosol-250-hhbr

7. National Library of Medicine, 2022, Sodium Bicarbonate. Retrieved March 9, 2022, from https://pubchem.ncbi.nlm.nih.gov.

8. Haz-Map, "Octoxynol, CAS# 9002-93-1", Haz-Map (2022, April 22), haz-map.com

9. https://chem.nlm.nih.gov/chemidplus/rn/9002-93-1

10. JimTrade, "Octylphenol Polyglycol Ether Sulphate Sodium Salt", Jim Trade (2022, April 22), https://www.jimtrade.com/

11. https://www.alfa.com/en/catalog/L15356/

12. National Library of Medicine, 2022, Butyl Acrylate. Retrieved March 9, 2022, from https://pubchem.ncbi.nlm.nih.gov.

13. Wikipedia, January 2022, Vinyl Acetate. Retrieved March 9, 2022, from https://en.wikipedia.org/

14. National Library of Medicine, 2022, Potassium Persulfate. Retrieved March 9, 2022, from https://pubchem.ncbi.nlm.nih.gov.

15. National Library of Medicine, 2022, Tert Butyl Hydroperoxide. Retrieved March 11, 2022, from https://pubchem.ncbi.nlm.nih.gov.

16. Nandi U. S., Palit S. R., Hydrogen peroxide as initiator in vinyl polymerization in homogeneous system. I. Kinetic studies, Journal of Polymer Science, 1955, 17 (83), pp 65-78.

17. National Library of Medicine, 2022, Sodium Formaldehyde Sulfoxylate. Retrieved March 12, 2022, fromhttps://pubchem.ncbi.nlm.nih.gov.

18. https://coatings.specialchem.com/product/a-bruggemannreducing- agent-tp-1648

19. Wikipedia, January 2022, Dibutyl Phthalate. Retrieved February 11, 2022, from https://en.wikipedia.org/

20. Stefano Ciroi, 2017, Formaldehyde in Vinyl Adhesives. Retrieved March 13, 2022, from https://catas.com/en- GB/news.

21. Michalak, K. Chojnacka, Biocides, Encyclopedia of Toxicology (Third Edition) 2014, Pages 461-463.

22. National Library of Medicine, 2022, Vinyltrimethoxysilane. Retrieved March 9, 2022, from https://pubchem.ncbi.nlm.nih.gov.

23. Momentive Solutions for a Sustainable Group, 2022, Silquest A-171 Silane. Retrieved March 12, 2022, from www.momentive.com.

24. Encyclopedia Britannica, Inc., 2022, Emulsion Polymerization, Retrieved 10 March 2022, from www.Britannica.com

25. Grechanovskii, V. A., Branching in Polymer Chains, *Rubber Chemistry and Technology* (1972) 45 (3): 519–545.

26. *Charles E. C Jr., 2017,* Introduction to Polymers 4th edition,Boca Raton, Taylor & Francis Group.

27. Swamy, R. N. and Tanikawa, S., 'An external surface coating to protect concrete and steel from aggressive environments', *Mater. Struct.* 26 (1993) 465-478.

28. Tyssal, L. A., 'Industrial paints: Basic principles', (Pergamon Press Ltd, London 1964).

29. Uemoto, K. L., Agopyan, V., Ranieri, R. and Quarcioni, V. A.,'Effect of concrete coating systems on chloride penetration', in 'Consec 98: Concrete Under Severe Conditions', Proceedings of an International Conference, Norway, (1998) 1321-1330.

30. Uemoto KL, Agopyan V, Vittorino F. Concrete protection using acrylic latex paints: Effect of the pigment volume content on water permeability. Materials and Structures. 2001 Apr; 34(3):172-7.

Chapter 3. Increase Product Quality for a Car-wash Shampoo Concentrate

1. www. sakagroup.com

2. Company M., Karsa D. R., Handbook for Cleaning/ Decontamination of Surfaces, Amsterdam, the Netherlands, Elsevier B.V., 2007.

3. H.G. Hauthal and G. Wagner (eds.), Household Cleaning, Care and Maintenance Products: Chemistry, Application, Ecology and Consumer Safety, publ. Verlag Fur chemische Industrie H Ziolkowsky GmbH, 2004.

4. Surfactants Selector; A Guide to the Selection of I&I and Household Product Formulations, Akcros Chemicals (now part of Akzo Nobel Surface Chemistry AB) 1998.

5. Further formulation information available from Akzo Nobel Surface Chemistry AB, S 444 85 Stenungsund, Sweden, on request.

6. Chemicalland21, 2013, Linear Alkylbenzene Sulfonic Acid, Retrieved May 2022, from www.chemicalland21.com/specialtychem/perchem.

7. Wikipedia the Free Encyclopedia, 2022, Sodium Laureth Sulfate, Retrieved May 2022, from https://en.wikipedia.org.

8. Wikipedia the Free Encyclopedia, 2022, Sodium Dodecyl Sulfate, Retrieved May 2022, from https://en.wikipedia.org.

9. Wikipedia the Free Encyclopedia, 2022, Sodium Hydroxide, Retrieved May 2022, from https://en.wikipedia.org.

10. Wikipedia the Free Encyclopedia, 2022, Trimethylglycine, Retrieved May 2022, from https://en.wikipedia.org.

11. Wikipedia the Free Encyclopedia, 2022, water, Retrieved May 2022, from https://en.wikipedia.org.

12. Wikipedia the Free Encyclopedia, 2022, Sodium Triphosphate, Retrieved May 2022, from https://en.wikipedia.org.

13. Wikipedia the Free Encyclopedia, 2022, Glycerol, Retrieved May 2022, from https://en.wikipedia.org.

14. Wikipedia the Free Encyclopedia, 2022, Propylene Glycol, Retrieved May 2022, from https://en.wikipedia.org.

15. World of Chemicals, 2022, Linear Alkylbenzene Sulfonic Acid, Retrieved May 2022, from www.worldofchemicals.com.

16. Valappil K., Lalitha S., Gottumukkala D., Sukumaran R. K., Pandey A., White Biotechnology in Cosmetics, Industrial Biorefineries & White Biotechnology, 2015, pp. 607-652. ISBN 9780444634535

17. "Water Q&A: Why is water the "universal solvent"?". www.usgs.gov. (U.S. Department of the Interior). Retrieved 15 January 2021.

18. "CIA – THE WORLD FACTBOOK Geography Geographic overview". Central Intelligence Agency. Retrieved 20 December 2008.

19. Baroni, L.; Cenci, L.; Tettamanti, M.; Berati, M. (2007). "Evaluating the environmental impact of various dietary patterns combined with different food production systems". European Journal of Clinical Nutrition. 61 (2): 279–286. doi:10.1038/sj.ejcn.1602522. PMID 17035955.

20. Complexing agents, Environmental and Health Assessment of Substances in Household Detergents and Cosmetic Detergent Products, Danish Environmental Protection Agency, Accessed 2008-07-15

21. Schrödter, Klaus; Bettermann, Gerhard; Staffel, Thomas; Wahl, Friedrich; Klein, Thomas; Hofmann, Thomas (2008). "Phosphoric Acid and Phosphates". Ullmann's Encyclopedia of Industrial Chemistry. doi:10.1002/14356007.a19_465.pub3. ISBN 978-3527306732.

22. Oxford Dictionaries – English, glycerol - Definition of glycerol in English by Oxford Dictionaries". Archived from the original on 21 June 2016. Retrieved 21 February 2022.

23. Christoph, Ralf; Schmidt, Bernd; Steinberner, Udo; Dilla, Wolfgang; Karinen, Reetta (2006). "Glycerol". Ullmann's Encyclopedia of Industrial Chemistry. doi:10.1002/14356007.a12_477.pub2. ISBN 3527306730.

24. Sullivan, Carl J.; Kuenz, Anja; Vorlop, Klaus-Dieter (2018). "Propanediols". Ullmann's Encyclopedia of Industrial Chemistry. Weinheim: Wiley-VCH. doi:10.1002/14356007.a22_163.pub2.

25. Lohrey, Jackie. "Ingredients in Hand Sanitizer". LIVESTRONG.COM. Retrieved 2018-06-11.

26. "Quackmail: Why You Shouldn't Fall For The Internet's Newest Fool, The Food Babe". Butterworth, Trevor. Forbes. 16 June 2014. Retrieved 18 March 2015.

27. G. Jackson, R. T. Roberts and T. Wainwright (January 1980). "Mechanism of Beer Foam Stabilization by Propylene Glycol Alginate". Journal of the Institute of Brewing. 86 (1): 34–37. doi:10.1002/j.2050-0416.1980.tb03953.x.

28. PersianUTab, "Sodium citrate in detergents", www. persianutab.com, 2020. Retrieved May 11, 2022

29. Contact Dermatitis Institute, "Coconut diethanolamide", www.contactdermatitisinstitute.com, 2022. Retrieved May 11, 2022.

Chapter 4. Modification to An Acrylic Paint Formulation

1. http://pscf.com.sa/

2. Izzo, Francesca Caterina; Balliana, Eleonora; Pinton, Federica; Zendri, Elisabetta, "A preliminary study of the composition of commercial oil, acrylic and vinyl paints and their behavior after accelerated ageing conditions", 2014, 353–369. doi:10.6092/ISSN.1973-9494/4753.

3. Abdel-Wahab H., Gund, M. Chemical Formulations for Acrylic Matt and Acrylic Gloss Paints. American Journal of Applied and Industrial Chemistry. Vol. 6, No. 1, 2022, pp. 13-19. doi: 10.11648/j.ajaic.20220601.13

4. Nandi U. S., Palit S. R., Hydrogen peroxide as initiator in vinyl polymerization in homogeneous system. I. Kinetic studies, Journal of Polymer Science, 1955, 17 (83), pp 65-78.

5. I. Michalak, K. Chojnacka, Biocides, Encyclopedia of Toxicology (Third Edition) 2014, Pages 461-463.

Chapter 5. Modification to a Chemical Formulation for Optimal Performance

1. https://en.wikipedia.org/wiki/Acid%E2%80%93base_reaction

2. https://www.molinstincts.com/formula/Cornstarch-cfml-CT1087471098.html

3. https://en.wikipedia.org/wiki/Sodium_chloride

4. https://en.wikipedia.org/wiki/Potassium_bitartrate

5. https://en.wikipedia.org/wiki/Sodium_bicarbonate

6. https://pubchem.ncbi.nlm.nih.gov/compound/Citric-acid

7. https://en.wikipedia.org/wiki/Betanin

8. Ebbing and Gammon, 2015, General Chemistry, Cengage, P.103

9. Ebbing and Gammon, 2015, General Chemistry, Cengage, P.108

10. Ebbing and Gammon, 2015, General Chemistry, Cengage, P.111

11. Brown, LeMay, Burstein, Murphy, Woodward, Stottzfus,, 2017, Chemistry the Central Science, Pearson, P.93

12. Brown, LeMay, Burstein, Murphy, Woodward, Stottzfus,, 2017, Chemistry the Central Science, Pearson, P.94-95

13. https://pubchem.ncbi.nlm.nih.gov/compound/Glycerol

14. 11. Brown, LeMay, Burstein, Murphy, Woodward, Stottzfus,, 2017, Chemistry the Central Science, Pearson, P. 102

15. 11. Brown, LeMay, Burstein, Murphy, Woodward, Stottzfus,, 2017, Chemistry the Central Science, Pearson, P.106

16. 11. Brown, LeMay, Burstein, Murphy, Woodward, Stottzfus,, 2017, Chemistry the Central Science, Pearson, P.703

17. 11. Brown, LeMay, Burstein, Murphy, Woodward, Stottzfus,, 2017, Chemistry the Central Science, Pearson, P.1048

18. Chegeni A, Babaeipour V, Fathollahi M, Hosseini SG. Development of a new formulation of air revitalization tablets in closed atmospheres using the Taguchi statistical method. International Journal of Environmental Science and Technology. 2021:13:1-6.

19. Chen KH, Miao YB, Shang CY, Huang TY, Yu YT, Yeh CN, Song HL, Chen CT, Mi FL, Lin KJ, Sung HW. A bubble bursting-mediated oral drug delivery system that enables concurrent delivery of lipophilic and hydrophilic chemotherapeutics for treating pancreatic tumors in rats. Biomaterials. 2020;255:120157.

Chapter 6. Thermodynamic Study and Oxidation Products for the Reaction of Methyl Ethyl Sulfide with Oxygen at Standard Conditions

1. Pillai S, Bozzelli JW. Computational study on structures, thermochemical properties, and bond energies of disulfide oxygen (S-SO)- bridged CH3SSOH and CH3SS(= O)H and radicals. *J Phys Org Chem.* 2012;25:475-485.

2. Hindiyarti L, Glarborg P, Marshall P. Reactions of SO3 with the O/H radical pool under combustion conditions. *J Phys Chem A.* 2007;111:3984-3991.

3. J. Cao , W. Wang , Y. Zhang , T. Zhang , J. Lv , C. Li , Computational study on the re- action of CH 3 SCH 2 CH 3 with OH radical: mechanism and enthalpy of formation, Theor. Chem. Acc. 129 (2011) 771–780.

4. S. Pillai , J.W. Bozzelli , Computational study on structures, thermochemical properties, and bond energies of disulfide oxygen (S-S-O)-bridged CH 3 SSOH and CH 3 SS(= O)H and radicals, J. Phys. Org. Chem. 25 (2012) 475–485.

5. G. Oksdath-Mansilla , A.B. Peñéñory , I. Barnes , P. Wiesen , M.A. Teruel , Pho- todegradation of (CH 3 CH 2) 2S and CH 3 CH 2 SCH 3 initiated by OH radicals at atmospheric pressure. Product yields and mechanism in NOx free air, Atmos. Environ. 55 (2012) 263–270 .

6. A.J. Hynes , P.H. Wine , D.H. Semmes ,Kinetics and mechanism of hydroxyl reac- tions with organic sulfides, J. Phys. Chem. 90 (1986) 4148–4156 .

7. G. Oksdath-Mansilla , A.B. Peñéñory , I. Barnes , P. Wiesen , M.A. Teruel , Pho- todegradation of (CH 3 CH 2) 2S and CH 3 CH 2 SCH 3 initiated by OH radicals at atmospheric pressure. Product yields and mechanism in NOx free air, Atmos. Environ. 55 (2012) 263–270 .

8. R. Atkinson , J. Arey , Atmospheric degradation of volatile organic compounds, Chem. Rev. 103 (2003) 4605–4638 .

9. G. Oksdath-Mansilla , A.B. Peñéñory , M. Albu , I. Barnes , P. Wiesen , M.A. Teruel , FTIR relative kinetic study of the reactions of CH 3 CH 2 SCH 2 CH 3 and CH 3 CH 2 SCH 3 with OH radicals and Cl atoms at atmospheric pressure, Chem. Phys. Lett. 477 (2009) 22–27 .

10. G. Song, J.Bozzelli, Reaction pathways, kinetics and thermochemistry of the chemically-activated and stabilized primary methyl radical of methyl ethyl sulfide, CH 3 CH 2 SCH 2 • , with 3 O 2 to CH 2 CH 3 SCH 2 OO •, Combustion and flame, 204(2019) 368–379.

11. Guanghui Song, Joseph Bozzelli, Hebah Abdel-Wahab. Thermochemistry and Reaction Kinetics of Secondary Ethyl Radical of Methyl Ethyl

12. Sulfide, CH3SCH•CH3, with 3O2 to CH2SCH(OO•)CH3. *American Journal of Physical Chemistry*. Vol. 10, No. 4, 2021, pp. 74-87. doi: 10.11648/j.ajpc.20211004.14

13. Guanghui Song, Joseph W Bozzelli, Reaction kinetics and thermochemistry of the chemically activated and stabilized primary ethyl radical of methyl ethyl sulfide, CH3SCH2CH2•, with O2 to CH3SCH2CH2OO•, Wiley, *Int J Chem Kinet.* 2019;51:618–633.

14. Gammon, Ebbing, General Chemistry 11th edition, Boston MA, Cengage Learning, 2017, pp 630-631.

15. Gammon, Ebbing, General Chemistry 11th edition, Boston MA, Cengage Learning, 2017, pp 608-610.

16. Gammon, Ebbing, General Chemistry 11th edition, Boston MA, Cengage Learning, 2017, pp 618-620.

17. Kotz T., Townsend T., Chemistry and Chemical Reactivity 9th edition, 2015, Cengage Learning, pp-696-704.

18. https://en.wikipedia.org/wiki/Radical_(chemistry).

19. Zheng X., Fisher E.M, Gouldin F.C., Bozzelli J.W., Pyrolysis and Oxidation of Ethyl Methyl Sulfide in a Flow Reactor, Combustion and Flame, 158 (2011) 1049–1058.

20. Song G, Bozzelli J, Abdel-Wahab H. Thermochemistry and Reaction Kinetics of Secondary Ethyl Radical of Methyl Ethyl Sulfide, CH3SCH• CH3, with 3 O2 to CH2SCH (OO•) CH3. American Journal of Physical Chemistry. 2021;10(4):74-87.

21. da Silva RJ, Mojica-Sánchez LC, Gorza FD, Pedro GC, Maciel BG, Ratkovski GP, da Rocha HD, do Nascimento KT, Medina-Llamas JC, Chávez-Guajardo AE, Alcaraz-Espinoza JJ. Kinetics and thermodynamic studies of Methyl Orange removal by polyvinylidene fluoride-PEDOT mats. Journal of Environmental Sciences. 2021 Feb 1;100:62-73.

Chapter 7. Enthalpy, Entropy and Heat Capacity of some Fluorinated Ethanol's and its Radicals at Different Temperatures

1. Wang H., Castillo A., Bozzelli J.W., Thermochemical properties enthalpy, entropy , and heat capacity of C1-4 fluorinated hydrocarbons . J. Phys. Chem. 2015

2. Wang H., Bozzelli J.W., Thermochemical properties and Bond Dissociation Energy for Fluorinated Methanol and fluorinated methyl hydroperoxides, . J. Phys. Chem. 2016

3. Wallington, T. J.; Schneider, W. F.; Worsnop, D. R.; Nielsen, O. J.; Sehested, J.; Debruyn, W. J.; Shorter, J. A. The Environmental Impact of CFC Replacements HFCs and HCFCs. Environ. Sci. Technol. 1994, 28, 320A–326A.

4. Schneider, W. F.; Wallington, T. J. Ab Initio Investigation of the Heats of Formation of Several Trifluoromethyl Compounds. J. Phys. Chem. 1993, 97, 12783–12788.

5. Frisch, M. J.; Trucks, G. W.; Schlegel, H. B.; Scuseria, G. E.; Robb, M. A.; Cheeseman, J. R.; Scalmani, G.; Barone, V.; Mennucci, B.; Petersson, G. A.; et al. Gaussian 09; Gaussian, Inc.: Wallingford, CT, 2009.

6. Wang, H.; Castillo, Á.; Bozzelli, J. W. Thermochemical Properties Enthalpy, Entropy, and Heat Capacity of C1−C4 Fluorinated Hydrocarbons: Fluorocarbon Group Additivity. J. Phys. Chem. A 2015, 119, 8202−8215.

7. NIST Computational Chemistry Comparison and Benchmark Database. NIST Standard Reference Database Number , Release 16a; Johnson, R. D., III, Ed.; NIST: Gaithersburg, MD, 2013.

8. Sheng, C. Elementary, Pressure Dependent Model for Combustion of C1, C2 and Nitrogen Containing Hydrocarbons: Operation of A Pilot Scale Incinerator and Model Comparison. Ph.D. dissertation ; New Jersey Institute of Technology, 2002.

9. Becke, A. D. Density-functional Thermochemistry. III. The Role of Exact Exchange. J. Chem. Phys. 1993, 98, 5648−5652.

10. Lay, T. H.; Krasnoperov, L. N.; Venanzi, C. A.; Bozzelli, J. W.; *Shokhirev*, N. V. Ab Initio Study of α-Chlorinated Ethyl Hydro510peroxides CH 3CH2OOH, CH3CHClOOH, and CH3CCl2OOH: Conformational Analysis, Internal Rotation Barriers, Vibrational Frequencies, and Thermodynamic Properties. J. Phys. Chem. 1996, 513100 , 8240−8249.

11. Lay, T. H.; Krasnoperov, L. N.; Venanzi, C. A.; Bozzelli, J. W.; *Shokhirev*, N. V. Ab Initio Study of α-Chlorinated Ethyl Hydro peroxides CH3CH2OOH, CH3CHClOOH, and CH3CCl2OOH: Conformational Analysis, Internal Rotation Barriers, Vibrational 762 Frequencies, and Thermodynamic Properties. J. Phys. Chem. 1996, 100, 8240−8249.

12. NIST Computational Chemsitry Comparison and Benchmark Database, NIST Standard Reference Database Number 101, Release 16a; Johnson, R. D. III, Ed.; NIST: Gaithersburg, MD, http://cccbdb. nist.gov (accessed Aug 2013).

13. Schneider, W. F.; Wallington, T. J. Ab Initio Investigation of the Heats of Formation of Several Trifluoromethyl Compounds. J. Phys. 420 Chem. 1993, 97, 12783−12788

14. Sawada, H. Fluorinated Peroxides. Chem. Rev. 1996, 96, 1779 1808.

15. Hayman, G. D.; Derwent, R. G. Atmospheric Chemical Reactivity and Ozone-Forming Potentials of Potential CFC Replacements. Environ. Sci. Technol. 1997, 31, 327−336.

16. El-Taher, S. Ab Initio and DFT Investigation of Fluorinated Methyl Hydroperoxides: Structures, Rotational Barriers, and Thermo chemical Properties. J. Fluorine Chem. 2006, 127, 54–62.

17. Reints, W.; Pratt, D. A.; Korth, H.-G.; Mulder, P. O–O Bond Dissociation Enthalpy in Di(trifluoromethyl) Peroxide (CF3OOCF3) as Determined by Very Low-Pressure Pyrolysis. Density Functional Theory Computations on O–O and O–H Bonds in (Fluorinated) Derivatives. J. Phys. Chem. A 2000, 104, 10713–10720.

18. Harvey, J.; Tuckett, R. P.; Bodi, A. A Halomethane Thermochemical Network from iPEPICO Experiments and Quantum Chemical Calculations. J. Phys. Chem. A 2012, 116, 9696–9705.

19. Kosmas, A. M.; Mpellos, C.; Salta, Z.; Drougas, E. Structural, and Heat of Formation Studies of Halogenated Methyl Hydro-peroxides. Chem. Phys. 2010, 371, 36–42.

20. Wang, H.; Bozzelli, J. W. Thermochemical Properties ($\Delta fH°(298$ K),S °(298 K),C p(T)) and Bond Dissociation Energies for C1–C4 Normal Hydroperoxides and Peroxy Radicals. J. Chem. Eng. Data 2016, 61, 1836–1849.

21. Wang, H.; Castillo, Á.; Bozzelli, J. W. Thermochemical Properties Enthalpy, Entropy, and Heat Capacity of C1–C4 Fluorinated Hydrocarbons: Fluorocarbon Group Additivity. J. Phys. Chem. A 2015, 119, 8202–8215.

22. Frisch, M. J.; Trucks, G. W.; Schlegel, H. B.; Scuseria, G. E.; Robb, M. A.; Cheeseman, J. R.; Scalmani, G.; Barone, V.; Mennucci, B.; Petersson, G. A.; et al. Gaussian 09; Gaussian, Inc.: Wallingford, CT, 2009.

23. Curtiss, L. A.; Redfern, P. C.; Raghavachari, K. Gaussian-4 Theory. J. Chem. Phys. 2007, 126, 084108.

24. Montgomery, J. A.; Frisch, M. J.; Ochterski, J. W.; Petersson, G. A. A Complete Basis Set Model Chemistry. VII. Use of the Minimum Population Localization Method. J. Chem. Phys. 2000, 112, 6532– 6542.

25. Ochterski, J. W.; Petersson, G. A.; Montgomery, J. A. A Complete Basis Set Model Chemistry. V. Extensions to Six or More Heavy Atoms. J. Chem. Phys. 1996, 104, 2598–2619.

26. Wang, H.; Bozzelli, J. W. Thermochemistry and Kinetic Analysis on Unimolecular Dissociation Reaction of Oxiranyl Radical: A Theoretical Study. ChemPhysChem 2016, 17, 1983–1992.

27. Ruscic, B. Active Thermochemical Tables: Sequential Bond Dissociation Enthalpies of Methane, Ethane, and Methanol and the Related Thermochemistry. J. Phys. Chem. A 2015, 119, 7810–7837.

28. Burke, S. M.; Simmie, J. M.; Curran, H. J. Critical Evaluation of Thermochemical Properties of C1 −C4 Species: Updated Group Contributions to Estimate Thermochemical Properties. J. Phys. Chem. Ref. Data 2015, 44, 013101.

29. Chase, M. W. J. NIST-JANAF Thermochemical Tables. J. Phys. Chem. Ref. Data. 1998, Monograph 9,1 −1951.

30. Luo, X.; Fleming, P. R.; Rizzo, T. R. Vibrational Overtone Spectroscopy of the 4 νOH+νOH′ Combination Level of HOOH via Sequential Local Mode −local Mode Excitation. J. Chem. Phys. 1992, 96 , 5659−5667.

31. Bodi, A.; Kercher, J. P.; Bond, C.; Meteesatien, P.; Sztáray, B.; Baer, T. Photoion Photoelectron Coincidence Spectroscopy of Primary Amines RCH 2NH2 (R = H, CH3,C 2H5,C 3H7, i-C3H7): Alkylamine and Alkyl Radical Heats of Formation by Isodesmic Reaction Networks. J. Phys. Chem. A 2006, 110, 13425−13433.

32. Csontos, J.; Rolik, Z.; Das, S.; Kállay, M. High-Accuracy Thermochemistry of Atmospherically Important Fluorinated and Chlorinated Methane Derivatives. J. Phys. Chem. A 2010, 114, 13093 −13103.

33. NIST Computational Chemistry Comparison and Benchmark Database. NIST Standard Reference Database Number 101, Release 16a; Johnson, R. D., III, Ed.; NIST: Gaithersburg, MD, 2013.

34. Becke, A. D. Density-functional Thermochemistry. III. The Role of Exact Exchange. J. Chem. Phys. 1993, 98, 5648−5652.

35. Sheng, C. Elementary, Pressure Dependent Model for Combustion of C1, C2 and Nitrogen Containing Hydrocarbons: Operation of A Pilot Scale Incinerator and Model Comparison. Ph.D. dissertation ; New Jersey Institute of Technology, 2002

36. Math is Fun Advanced, 2017, *Standard Deviation and Variance*, 12/2019,{https://www.mathsisfun.com/data/standard-deviation.html}

37. Ruscic ,B., Active Thermochemical Tables: Sequential Bond Dissociation Enthalpy of Methane, Ethane, and Methanol and related Thermochemistry. J. Phys. Chem. A 2015,119, 7810-7837

38. Argonne National Laboratory, 2019, *Active Thermochemical Tables,* US. Department of Energy, 10/26/2017, 'https://atct.anl.gov/Thermochemical%20Data/version%201.118/index.p hp"

39. Myrna H. M., Minh T. N., David A D, Minh T. N., Theoretical Prediction of the Heats of Formation of C2H5O* Radicals Derived From Ethanol and

of the Kinetics of beta-C-C Scission in the Ethoxy Radical, J. Phys. Chem. A 2007, 111, 1, 113–126

40. National Computational Science Institute,1994-2020, *Computational Chemistry Methods*, 7/25/2020, 'http://www.computationalscience.org/ccce/Lesson2/Notebook%202%20 Lecture.pdf,'

41. Expanding The Limits of Computational Chemistry, 2015-2019, *Density Function Method,* 7/24/2020, 'https://gaussian.com/dft/'

42. Bao G, Abe RY, Akutsu Y. Bond dissociation energy and thermal stability of energetic materials. Journal of Thermal Analysis and Calorimetry. 2021143(5):3439-45.

43. Sorensen JJ, Tieu E, Morse MD. Bond dissociation energies of lanthanide sulfides and selenides. The Journal of Chemical Physics. 2021;154(12):124307.

Chapter 8. Thermochemistry and Reaction Kinetics of the Stabilized and the Chemically Activated Secondary Ethyl Radical of Methyl Ethyl Sulfide with Oxygen

1. K. Vijayaraghavan, K. Hemanathan, Biodiesel production from freshwater algae, Energy and Fuels 23 (2009) 5448-5453.

2. F.G. Cerru, A. Kronenburg, R.P. Lindstedt, Systematically reduced chemical mechanisms for sulfur oxidation and pyrolysis, Combustion and Flame 146 (2006) 437-455.

3. L. Hindiyarti, P. Glarborg, P. Marshall, Reactions of so$_3$ with the O/H radical pool under combustion conditions, Journal of Physical Chemistry A 111 (2007) 3984-3991.

4. I.A. Gargurevich, Hydrogen sulfide combustion: Relevant issues under claus furnace conditions, Industrial and Engineering Chemistry Research 44 (2005) 7706-7729.

5. A. Gross, I. Barnes, R.M. Sørensen, J. Kongsted, K.V. Mikkelsen, A Theoretical Study of the Reaction between CH3S(OH)CH3 and O2, The Journal of Physical Chemistry A 108 (2004) 8659-8671.

6. M.B. Williams, P. Campuzano-Jost, A.J. Pounds, A.J. Hynes, Experimental and theoretical studies of the reaction of the OH radical with alkyl sulfides: 2. Kinetics and mechanism of the OH initiated oxidation of methylethyl and diethyl sulfides; observations of a two channel oxidation mechanism, Physical Chemistry Chemical Physics 9 (2007) 4370-4382.

7. M.B. Williams, P. Campuzano-Jost, A.J. Hynes, A.J. Pounds, Experimental and theoretical studies of the reaction of the OH radical with alkyl sulfides: 3. Kinetics and mechanism of the OH initiated oxidation of dimethyl, dipropyl, and dibutyl sulfides: reactivity trends in the alkyl sulfides and development of a predictive expression for the reaction of OH with DMS, Journal of Physical Chemistry A 113 (2009) 6697-6709.

8. J. Cao, W. Wang, Y. Zhang, T. Zhang, J. Lv, C. Li, Computational study on the reaction of CH3SCH2CH3 with OH radical: Mechanism and enthalpy of formation, Theoretical Chemistry Accounts 129 (2011) 771-780.

9. S. Pillai, J.W. Bozzelli, Computational study on structures, thermochemical properties, and bond energies of disulfide oxygen (S-S-O)-bridged CH 3SSOH and CH 3SS(=O)H and radicals, Journal of Physical Organic Chemistry 25 (2012) 475-485.

10. C.S. Turchi, D.F. Ollis, Photocatalytic degradation of organic water contaminants: Mechanisms involving hydroxyl radical attack, Journal of Catalysis 122 (1990) 178-192.

11. D. Grosjean, Photooxidation of methyl sulfide, ethyl sulfide, and methanethiol, Environmental Science and Technology 18 (1984) 460-468.

12. A.J. Hynes, P.H. Wine, D.H. Semmes, Kinetics and mechanism of OH reactions with organic sulfides, Journal of Physical Chemistry 90 (1986) 4148-4156.

13. A.J. Hynes, P.H. Wine, D.H. Semmes, Kinetics and mechanism of hydroxyl reactions with organic sulfides, The Journal of Physical Chemistry 90 (1986) 4148-4156.

14. M.B. Williams, P. Campuzano-Jost, B.M. Cossairt, A.J. Hynes, A.J. Pounds, Experimental and theoretical studies of the reaction of the OH radical with alkyl sulfides: 1. Direct observations of the formation of the OH-DMS adduct-pressure dependence of the forward rate of addition and development of a predictive expression at low temperature, Journal of Physical Chemistry A 111 (2007) 89-104.

15. S.P. Pillai, Structure and Thermochemistry of Disulfide-Oxygen Species, Department of Chemistry and Environmental Science, New Jersey Institute of Technology, Newark, NJ, 2008.

16. A.G. Vandeputte, M.-F. Reyniers, G.B. Marin, Theoretical Study of the Thermal Decomposition of Dimethyl Disulfide, Journal of Physical Chemistry A 114 (2010) 10531-10549.

17. X. Zheng, J.W. Bozzelli, E.M. Fisher, F.C. Gouldin, L. Zhu, Experimental and computational study of oxidation of diethyl sulfide in a flow reactor, Proceedings of the Combustion Institute 33 (2011) 467-475.

18. G. Oksdath-Mansilla, A.B. Peñéñory, I. Barnes, P. Wiesen, M.A. Teruel, Photodegradation of (CH 3CH 2) 2S and CH 3CH 2SCH 3 initiated by OH radicals at atmospheric pressure. Product yields and mechanism in NOx free air, Atmospheric Environment 55 (2012) 263-270.

19. J.A. Pople, M. Head-Gordon, D.J. Fox, K. Raghavachari, L.A. Curtiss, Gaussian-1 theory: A general procedure for prediction of molecular energies, The Journal of Chemical Physics 90 (1989) 5622-5629.

20. J.A. Pople, Nobel lecture: Quantum chemical models, Reviews of Modern Physics 71 (1999) 1267-1274.

21. M.J. Frisch, G.W. Trucks, H.B. Schlegel, G.E. Scuseria, M.A. Robb, J.R. Cheeseman, J.A. Montgomery, T. Vreven, K.N. Kudin, J.C. Burant, J.M. Millam, S.S. Iyengar, J. Tomasi, V. Barone, B. Mennucci, M. Cossi, G. Scalmani, N. Rega, G.A. Petersson, H. Nakatsuji, M. Hada, M. Ehara, K. Toyota, R. Fukuda, J. Hasegawa, M. Ishida, T. Nakajima, Y. Honda, O. Kitao, H. Nakai, M. Klene, X. Li, J.E. Knox, H.P. Hratchian, J.B. Cross, V. Bakken, C. Adamo, J. Jaramillo, R. Gomperts, R.E. Stratmann, O. Yazyev, A.J. Austin, R. Cammi, C. Pomelli, J.W. Ochterski, P.Y. Ayala, K. Morokuma, G.A. Voth, P. Salvador, J.J. Dannenberg, V.G. Zakrzewski, S. Dapprich, A.D. Daniels, M.C. Strain, O. Farkas, D.K. Malick, A.D. Rabuck, K. Raghavachari, J.B. Foresman, J.V. Ortiz, Q. Cui, A.G. Baboul, S. Clifford, J. Cioslowski, B.B. Stefanov, G. Liu, A. Liashenko, P. Piskorz, I. Komaromi, R.L. Martin, D.J. Fox, T. Keith, A. Laham, C.Y. Peng, A. Nanayakkara, M. Challacombe, P.M.W. Gill, B. Johnson, W. Chen, M.W. Wong, C. Gonzalez, J.A. Pople, Gaussian 03, Revision C.02, 2003.

22. J.A. Montgomery Jr, M.J. Frisch, J.W. Ochterski, G.A. Petersson, A complete basis set model chemistry. VII. Use of the minimum population localization method, Journal of Chemical Physics 112 (2000) 6532-6542.

23. J.A. Montgomery Jr, M.J. Frisch, J.W. Ochterski, G.A. Petersson, A complete basis set model chemistry. VI. Use of density functional geometries and frequencies, Journal of Chemical Physics 110 (1999) 2822-2827.

24. L.A. Curtiss, P.C. Redfern, K. Raghavachari, V. Rassolov, J.A. Pople, Gaussian-3 theory using reduced Møller-Plesset order, Journal of Chemical Physics 110 (1999) 4703-4709.

25. A.G. Baboul, L.A. Curtiss, P.C. Redfern, K. Raghavachari, Gaussian-3 theory using density functional geometries and zero-point energies, Journal of Chemical Physics 110 (1999) 7650-7657.

26. L.A. Curtiss, K. Raghavach Ari, P.C. Redfern, V. Rassolov, J.A. Pople, Gaussian-3 (G3) theory for molecules containing first and second-row atoms, Journal of Chemical Physics 109 (1998) 7764-7776.

27. K.B. Wiberg, G.A. Petersson, A Computational Study of RXHn X-H Bond Dissociation Enthalpies, Journal of Physical Chemistry A 118 (2014) 2353-2359.

28. K.B. Wiberg, G.B. Ellison, J.M. McBride, G.A. Petersson, Substituent Effects on O-H Bond Dissociation Enthalpies: A Computational Study, Journal of Physical Chemistry A 117 (2013) 213-218.

29. V.G. Kiselev, N.P. Gritsan, V.E. Zarko, P.I. Kalmykov, V.A. Shandakov, Multilevel quantum chemical calculation of the enthalpy of formation of 1,2,5 oxadiazolo 3,4-e 1,2,3,4 -tetrazine-4,6-Di-N-Dioxide, Combustion Explosion and Shock Waves 43 (2007) 562-566.

30. C.Y. Sheng, J.W. Bozzelli, A.M. Dean, A.Y. Chang, Detailed Kinetics and Thermochemistry of C2H5 + O2: Reaction Kinetics of the Chemically-Activated and Stabilized CH3CH2OO• Adduct, The Journal of Physical Chemistry A 106 (2002) 7276-7293.

31. K.B. Wiberg, Ab Initio Molecular Orbital Theory, J. Comput. Chem. 7 (1986) 379-379.

32. I.K. Ortega, O. Kupiainen, T. Kurtén, T. Olenius, O. Wilkman, M.J. McGrath, V. Loukonen, H. Vehkamäki, From quantum chemical formation free energies to evaporation rates, Atmospheric Chemistry and Physics 12 (2012) 225-235.

33. R. Casasnovas, J. Frau, J. Ortega-Castro, A. Salvà, J. Donoso, F. Muñoz, Simplification of the CBS-QB3 method for predicting gas-phase deprotonation free energies, International Journal of Quantum Chemistry 110 (2010) 323-330.

34. A.G. Vandeputte, M.K. Sabbe, M.F. Reyniers, G.B. Marin, Modeling the Gas-Phase Thermochemistry of Organosulfur Compounds, Chemistry-a European Journal 17 (2011) 7656-7673.

35. C.A. Class, J. Aguilera-Iparraguirre, W.H. Green, A kinetic and thermochemical database for organic sulfur and oxygen compounds, Physical Chemistry Chemical Physics 17 (2015) 13625-13639.

36. A.D. Becke, Density-functional thermochemistry. III. The role of exact exchange, The Journal of Chemical Physics 98 (1993) 5648-5652.

37. T.H. Lay, L.N. Krasnoperov, C.A. Venanzi, J.W. Bozzelli, N.V. Shokhirev, Ab initio study of α-chlorinated ethyl hydroperoxides CH3CH2OOH, CH3CHClOOH, and CH3CCl2OOH: Conformational analysis, internal rotation barriers, vibrational frequencies, and thermodynamic properties, Journal of Physical Chemistry 100 (1996) 8240-8249.

38. H. Wang, Á. Castillo, J.W. Bozzelli, Thermochemical Properties Enthalpy, Entropy, and Heat Capacity of C1–C4 Fluorinated Hydrocarbons: Fluorocarbon Group Additivity, The Journal of Physical Chemistry A 119 (2015) 8202-8215.

39. B.A. Ellingson, V.A. Lynch, S.L. Mielke, D.G. Truhlar, Statistical thermodynamics of bond torsional modes: Tests of separable, almost-separable, and improved Pitzer-Gwinn approximations, Journal of Chemical Physics 125 (2006).

40. P. Hui-Yun, The modified Pitzer-Gwinn method for partition function of restricted internal rotation of a molecule, The Journal of Chemical Physics 87 (1987) 4846-4848.

41. K.S. Pitzer, Thermodynamic functions for molecules having restricted internal rotations, The Journal of Chemical Physics 5 (1937) 469-472.

42. G. da Silva, J.W. Bozzelli, Enthalpies of formation, bond dissociation energies, and molecular structures of the n-aldehydes (acetaldehyde, propanal, butanal, pentanal, hexanal, and heptanal) and their radicals, Journal of Physical Chemistry A 110 (2006) 13058-13067.

43. S.R. Hashemi, V. Saheb, Theoretical studies on the mechanism and kinetics of the hydrogen abstraction reactions of threo-CF3CHFCHFC2F5 and erythro-CF3CHFCHFC2F5 (HFC-43-10mee) by OH radicals, Computational and Theoretical Chemistry 1119 (2017) 59-64.

44. H.S. Johnston, J. Heicklen, TUNNELLING CORRECTIONS FOR UNSYMMETRICAL ECKART POTENTIAL ENERGY BARRIERS, The Journal of Physical Chemistry 66 (1962) 532-533.

45. H.S. Johnston, Gas phase reaction rate theory, New York, Ronald Press Co. [1966]1966.

46. S. Yommee, J.W. Bozzelli, Cyclopentadienone Oxidation Reaction Kinetics and Thermochemistry for the Alcohols, Hydroperoxides, and Vinylic, Alkoxy, and Alkylperoxy Radicals, Journal of Physical Chemistry A 120 (2016) 433-451.

47. F. Jin, R. Asatryan, J.W. Bozzelli, Thermodynamic and kinetic analysis on the reaction of dimethyl sulfide radical with oxygen, International Journal of Quantum Chemistry 112 (2012) 1945-1958.

48. S. Snitsiriwat, J.W. Bozzelli, Thermochemistry, reaction paths, and kinetics on the tert -isooctane radical reaction with O2, Journal of Physical Chemistry A 118 (2014) 4631-4646.

49. H. Wang, J.W. Bozzelli, Thermochemistry and Kinetic Analysis of the Unimolecular Oxiranyl Radical Dissociation Reaction: A Theoretical Study, ChemPhysChem, doi:10.1002/cphc.201600152(2016).

50. B.E. Poling, J.M. Prausnitz, J.P. O'Connell, The properties of gases and liquids, New York : McGraw-Hill, c2001. 5th ed.2001.

51. R. Atkinson, J. Arey, Atmospheric Degradation of Volatile Organic Compounds, Chemical Reviews 103 (2003) 4605-4638.

52. G. Oksdath-Mansilla, A.B. Peñéñory, M. Albu, I. Barnes, P. Wiesen, M.A. Teruel, FTIR relative kinetic study of the reactions of CH3CH2SCH2CH3 and CH3CH2SCH3 with OH radicals and Cl atoms at atmospheric pressure, Chemical Physics Letters 477 (2009) 22-27.

53. L. Zhu, J.W. Bozzelli, L.M. Kardos, Thermochemical properties, Δ fH o(298), S o(298), and C p o(298), (T), for n-butyl and n-pentyl hydroperoxides and the alkyl and peroxy radicals, transition states, and kinetics for intramolecular hydrogen shift reactions of the peroxy radicals, Journal of Physical Chemistry A 111 (2007) 6361-6377.

54. G. Song, J.W. Bozzelli, Structures and thermochemistry of methyl ethyl sulfide and its hydroperoxides: HOOCH2SCH2CH3, CH3SCH(OOH)CH3, CH3SCH2CH2OOH, and radicals, Journal of Physical Organic Chemistry, doi:10.1002/poc.3751 e3751-n/a.

55. G. Song, J.W. Bozzelli, Structural and thermochemical studies on CH3SCH2CHO, CH3CH2SCHO, CH3SC(=O)CH3, and radicals corresponding to loss of H atom, Journal of Physical Organic Chemistry, doi:10.1002/poc.3688(2017) e3688-n/a.

56. B. Ruscic, Active Thermochemical Tables: Sequential Bond Dissociation Enthalpies of Methane, Ethane, and Methanol and the Related Thermochemistry, Journal of Physical Chemistry A 119 (2015) 7810-7837.

57. C.F. Goldsmith, G.R. Magoon, W.H. Green, Database of small molecule thermochemistry for combustion, Journal of Physical Chemistry A 116 (2012) 9033-9047.

58. Z. Xin, E.M. Fisher, F.C. Gouldin, J.W. Bozzelli, Pyrolysis and oxidation of ethyl methyl sulfide in a flow reactor, Combustion and Flame 158 (2011) 1049-1058.

59. X. Zheng, E.M. Fisher, F.C. Gouldin, L. Zhu, J.W. Bozzelli, Experimental and computational study of diethyl sulfide pyrolysis and mechanism, Proceedings of the Combustion Institute 32 (2009) 469-476.

60. K.P. Somers, J.M. Simmie, F. Gillespie, C. Conroy, G. Black, W.K. Metcalfe, F. Battin-Leclerc, P. Dirrenberger, O. Herbinet, P.-A. Glaude, P. Dagaut, C. Togbé, K. Yasunaga, R.X. Fernandes, C. Lee, R. Tripathi, H.J. Curran, A comprehensive experimental and detailed chemical kinetic modelling study of 2,5-dimethylfuran pyrolysis and oxidation, Combustion and Flame 160 (2013) 2291-2318.

61. Genies, E.M., Boyle, A., Lapdowski, M., Tsintavis, C. Synth. Met. 1990, 36, 139-182.

62. Fu Y, Elsenbaumer RL. Thermochemistry and kinetics of chemical polymerization of aniline determined by solution calorimetry. Chemistry of materials. 1994 May 1;6(5):671-7.

Chapter 9. Enthalpy and Bond dissociation Energy Values for Multi-Fluorinated Ethanol's and its Radicals using Gaussian M-062x/6-31+g (d,p) Method at Standard Conditions.

1. An X., Xiao J., Fluorinated Alcohols: Magic Reaction Medium and Promoters for Organic Synthesis. The Chemical record, a journal of the chemical society of Japan. 2020, 20(2): 142-161.

2. Zhang M., Peyear T., Patmanidis I.,Greathouse D.V., Marrink S.J., Andersen O.S., and Ingólfsson H.I. Fluorinated Alcohols' Effects on Lipid Bilayer Properties. Biophysical Journal. 2018, 115(4): 679–689.

3. J. P.Ducrotoy, K.Mazik. Chemical Introductions to the Systems: Point Source Pollution (Persistent Chemicals). Treatise on Estuarine and Coastal Science. 2011, Volume 8: 71-111

4. Bao G, Abe RY, Akutsu Y. Bond dissociation energy and thermal stability of energetic materials. Journal of Thermal Analysis and Calorimetry. 2021 Mar;143(5):3439-45.

5. Sorensen JJ, Tieu E, Morse MD. Bond dissociation energies of lanthanide sulfides and selenides. The Journal of Chemical Physics. 2021 Mar 28;154(12):124307.

6. https://chem.libretexts.org/Bookshelves/Organic_Chemistry/Organic_Ch emistry_(McMurry)/06%3A_An_Overview_of_Organic_Reactions/6.08 %3A_Describing_a_Reaction__Bond__Dissociation_Energies.

7. https://chem.libretexts.org/Courses/Heartland_Community_College/HCC%3A_Chem_161/5%3A_Thermochemistry/5.7%3A_Enthalpy_of_Formation

8. Gammon, Ebbing, General Chemistry 11th edition, Boston MA, Cengage Learning, 2017, pp 608-610. https://en.wikipedia.org/wiki/Radical_(chemistry)

9. Wang H., Castillo A., Bozzelli J.W., Thermochemical properties enthalpy, entropy , and heat capacity of C1-4 fluorinated hydrocarbons . J. Phys. Chem. 2015

10. Schneider, W. F.; Wallington, T. J. Ab Initio Investigation of the Heats of Formation of Several Trifluoromethyl Compounds. J. Phys. Chem. 1993, 97, 12783−12788.

11. Expanding The Limits of Computational Chemistry, 2015-2019, *Density Function Method,* 7/24/2020, 'https://gaussian.com/dft/.

12. Wang, H.; Castillo, Á.; Bozzelli, J. W. Thermochemical Properties Enthalpy, Entropy, and Heat Capacity of C1−C4 Fluorinated Hydrocarbons: Fluorocarbon Group Additivity. J. Phys. Chem. A 2015, 119, 8202−8215.

13. Math is Fun Advanced, 2017, *Standard Deviation and Variance,* 12/2019,{https://www.mathsisfun.com/data/standard-deviation.html

14. Wang H., Bozzelli J. W., Thermochemical properties and Bond Dissociation Energy for Fluorinated Methanol and fluorinated methyl hydroperoxides,. J. Phys. Chem. 2016.

15. Ruscic, B., Active Thermochemical Tables: Sequential Bond Dissociation Enthalpy of Methane, Ethane, and Methanol and related Thermochemistry. J. Phys. Chem. A 2015, 119, 7810-7837.

16. Burke, S. M.; Simmie, J. M.; Curran, H. J. Critical Evaluation of Thermochemical Properties of C1 −C4 Species: Updated Group Contributions to Estimate Thermochemical Properties. J. Phys. Chem. Ref. Data 2015, 44, 013101.

17. Chase, M. W. J. NIST-JANAF Thermochemical Tables. J. Phys. Chem. Ref. Data. 1998, Monograph 9, 1-1951.

18. Luo, X.; Fleming, P. R.; Rizzo, T. R. Vibrational Overtone Spectroscopy of the 4 vOH+vOH′ Combination Level of HOOH via Sequential Local Mode −local Mode Excitation. J. Chem. Phys. 1992, 96, 5659−5667.

19. Bodi, A.; Kercher, J. P.; Bond, C.; Meteesatien, P.; Sztaray, B.; Baer, T. Photoion Photoelectron Coincidence Spectroscopy of Primary Amines RCH 2NH2 (R = H, CH3, C 2H5, C 3H7, i-

20. C3H7): Alkylamine and Alkyl Radical Heats of Formation by Isodesmic Reaction Networks. J. Phys. Chem. A 2006, 110, 13425−13433.

21. Wang, H.; Bozzelli, J. W. Thermochemical Properties (ΔfH (298 K), S (298 K), C p(T)) and Bond Dissociation Energies for C1−C4 Normal Hydroperoxides and Peroxy Radicals. J. Chem. Eng. Data 2016, 61, 1836−1849.

22. Csontos, J.; Rolik, Z.; Das, S.; Kallay, M. High-Accuracy Thermochemistry of Atmospherically Important Fluorinated and Chlorinated Methane Derivatives. J. Phys. Chem. A 2010, 114, 13093 −13103.

23. Gronert, S. *J. Org. Chem.* 2006, *13*, 1209.

Chapter 10. Hypoxemia and hypercapnia: its Relation to Joint Diseases and The Natural Treatments for Some Joint Diseases. A Review

1. Global Climate Change, "10 interesting things about air ", September 2016, www.climate.nasa.gov. Retrieved May 2022.

2. GDS Team, "Understanding Safe Oxygen Levels as Outlined by OSHA in Confined Spaces", May 2017, www.gdscorp.com. Retrieved May 2022.

3. Home Particle, "How is Oxygen level maintained in Air-Conditioned Room", November 2021, www.homeparticle.com. Retrieved May 2022.

4. Brennan D, "What's Respiratory Acidosis?", *www.webmd.com,* June 2021. Retrieved May 2022

5. US National Library of Medicine, "The Physiological Impact of N95 Masks on Medical Staff", *www. clinicaltrials.gov*, 2005. Retrieved May 31, 2022.

6. Contreras M, Masterson C, Laffey JG. Permissive hypercapnia: what to remember. Curr Opin Anaesthesiol. 2015 Feb;28(1):26-37. [PubMed]

7. Medical NEWS Today, "What's dyspnea?", www.medicalnewstoday.com, Medical News Today, 2018. Retrieved May 2022.

8. M.J. Sheild, "Anti-Inflammatory Drugs and their Effects on Catlilage Synthesis and Renal Function", Eur J Rheumatol Inflam 13 (1993):7-16.

9. 9. P.M. Brooks, S.R. Potter, and W.W. Buchanan, " NSAID and Osteoarthritis—Help or Hinderance", J Rheumatol 9 (1982): 3-5.

10. N.M Newman, and R.S.M. Ling, " Acetabular Bone Destruction Related to Non-steroidal Anti-Inflammatory Drugs", Lancet 2 (1985): 11-13.

11. L. Solomon, " Drug-Induced Anthropology and Necrosis of the Femoral Head", J Bone Joint Surg 55B (1973): 246-51.

12. H. Ronningen, and N. Langeland, " Indomethacin Treatment in Osteoarthritis of Hip Joints", Acta Orthop Scand 50(1979): 169-74.

13. Lisa Esposito, "A Patient's Guide to Bone and Joint Diseases", www.health.usnews.com, U.S. News, June 2019. Retrieved May 2022.

   A. Pristas, "Blood Oxygen Level: What's all the Hype about?" https://www.hackensackmeridianhealth.org/, September 2002. Retrieved May 2022.

14. Stanfield, C. L., Principles of Human Physiology 5th edition. Pearson, October 2012.

15. *Henry, W. (1803). "Experiments on the quantity of gases absorbed by water, at different temperatures, and under different pressures". Phil. Trans. R. Soc. Lond. 93: 29–43. doi:10.1098/rstl.1803.0004.*

16. Brinkman JE, Toro F, Sharma S. StatPearls [Internet]. StatPearls Publishing; Treasure Island (FL): Aug 24, 2021. Physiology, Respiratory Drive. [PubMed]

17. Kisaka T, Cox TA, Dumitrescu D, Wasserman K. CO2 pulse and acid-base status during increasing work rate exercise in health and disease. Respir Physiol Neurobiol. 2015 Nov;218:46-56. [PubMed]

18. Katalinić L, Blaslov K, Pasini E, Kes P, Bašić-Jukić N. [Acid-base status in patients treated with peritoneal dialysis]. Acta Med Croatica. 2014 Apr;68(2):85-90. [PubMed]

19. Marhong J, Fan E. Carbon dioxide in the critically ill: too much or too little of a good thing? Respir Care. 2014 Oct;59(10):1597-605. [PubMed]

20. Cove ME, Federspiel WJ. Veno-venous extracorporeal CO2 removal for the treatment of severe respiratory acidosis. Crit Care. 2015 Apr 17;19:176. [PMC free article] [PubMed]

21. Sharma S, Hashmi MF, Burns B. StatPearls [Internet]. StatPearls Publishing; Treasure Island (FL): Aug 30, 2021. Alveolar Gas Equation. [PubMed]

22. MacKenzie Burger and Derek J. Schalle, "Metabolic Acidolysis", National Library of Medicine, www.ncbi.nlm.nih.gov, 2021. Retrieved May 2022.

23. Expert Board, "Acidolysis and Alkalosis", www.testing.com, January 2022. Retrieved May 2022.

24. J. G. Betts, K. A. Young, J.A Wise, E. Johnson, B. Poe, D. H. Kruse, O. Korol, J.E. Johnson, M. Wombe, and P. Desaix, 2019, Anatomy and Sociology, OpenStax, http://cnx.org/contents/14fb4ad7-39a1-4eee-ab6e-3ef2482e3e22@15.5.

25. J. G. Betts, K. A. Young, J.A Wise, E. Johnson, B. Poe, D. H. Kruse, O. Korol, J.E. Johnson, M. Wombe, and P. Desaix, 2019, Anatomy and Sociology, OpenStax, http://cnx.org/contents/14fb4ad7-39a1-4eee-ab6e-3ef2482e3e22@15.5.

26. Quiñonez-Flores, C. M., González-Chávez, S. A., Pacheco-Tena, C., Hypoxia and its implication to Rheumatoid Arthritis, J Biomed Sci. 2016; 23(1): 62, doi: 10.1186/s12929-016-0281-0

27. 28. Zahan O.M., Serban O., Gherman C., Fodor D, The evaluation oxidative stress in osteoarthritis, Med Pharm Rep. 2020 Jan; 93(1): 12–22.

28. Neogi T, Chen C, Niu J, et al. Relation of temperature and humidity to the risk of recurrent gout attacks. Am J Epidemiol. 2014;180(4):372-377. DOI: 10.1093/aje/kwu147

29. Roddy E. Revisiting the pathogenesis of podagra: why does gout target the foot? J Foot Ankle Res. 2011;4(1):13. Published 2011 May 13. PMID: 21569453 doi:10.1186/1757-1146-4-13

30. Martillo MA, Nazzal L, Crittenden DB. The crystallization of monosodium urate. Curr Rheumatol Rep. 2014;16(2):400. PMID: 24357445 doi:10.1007/s11926-013-0400-9

31. .Choi HK, Niu J, Neogi T, Chen CA, Chaisson C, Hunter D, Zhang Y. Nocturnal risk of gout attacks. Arthritis Rheumatol. 2015 Feb;67(2):555-62. PMCID: PMC4360969 DOI: 10.1002/art.38917

32. 33. Abhishek A, Valdes AM, Jenkins W, Zhang W, Doherty M. Triggers of acute attacks of gout, does age of gout onset matter? A primary care-based cross-sectional study. PLoS One. 2017;12(10):e0186096. Published 2017 Oct 12. PMID: 29023487 doi:10.1371/journal.pone.0186096

33. Bouloukaki I, et al. (2014). Intensive versus standard follow-up to improve continuous positive airway pressure compliance. European Respiratory Journal, 44(5): 1262–1274. PMID: 24993911 DOI: 10.1183/09031936.00021314.

34. 35.R. Jenkins, P. Rooney, and D. Jones et al., "Increased Intestinal Permeability for Patients with Rheumatoid Arthritis: A Side Effect of Oral Non-Steroidal Anti-Inflammatory Drug Therapy ", Br J Rheumatol 26 (1987): 103-7.

35. Murray M, Pizzorno J., Encyclopedia of Natural Medicine 2nd edition, 1998. ISBN: 978-0-7615-1157-1

36. L. G. Darlington and N.W. Ramsey, "Clinical Review: Review of Dietary Therapy in Rheumatoid Arthritis ", Br J Rheumatol 30 (1993): 507-14.

37. H. M. Buchanan, S.J. Preston, and P.M. Brooks et al., " Is Diet Important in Rheumatoid Arthritis ", Br J Rheumatol 30(1991): 125-34.

38. F. McCrae, K. Veerapen, and P. Dieppe, "Diet and Arthritis ", Practitioner 230 (1986): 359-61.

39. T. J. De Witte et al., "Hypochlorhydria and Hypergastrinemia in Rheumatoid Arthritis ", Ann Rheu Dis 38 (1979): 14-17.

40. K. Henriksson et al., "Gastrin, Gastric Acid Secretion, and Gastric Microflora in Patients with Rheumatoid Arthritis ", Ann Rheu Dis 45 (1986): 475-83.

41. M. Brzeski, R. Madhok, H.A. Capell, "Evening Primrose Oil in Patients with Rheumatoid Arthritis and Side Effects on non-Steroidal Ant-Inflammatory Drugs ", Br J Rheumatol 30 (1991): 371-2.

42. J. F. Blech et al., "Effects of Altering Dietary Essential Fatty Acids on Requirements for non- steroidal Anti-inflammatory Drugs in Patients with Rheumatoid Arthritis: A Double-Blinded Placebo-Controlled Study ", Ann Rheu Dis 47 (1988): 96-104.

43. L. J. Levanthal et al.," Treatment of Rheumatoid Arthritis with Gamma-Linolenic Acid ", Annals Int Med 119 (1993): 867-73.

44. J. M. Kremer et al., "Fish Oil Supplementation in Active Rheumatoid Arthritis: A Double-Blinded, Controlled Cross-Over Study ", Ann Intern Med 106 (1987): 497-502.

45. R. Sperling et al., "Effects of Dietary Supplementation with Marine Fish Oil on Leukocyte Lipid Mediator Generation and Function in Rheumatoid Arthritis ", Arthritis Rheum 30 (1987): 988-97.

46. L. G. Cleland et al.," Clinical and Biochemical Effects of Dietary Fish Oil Supplementation in Rheumatoid Arthritis ", J Rheumatol 15 (1988): 1471-5.

47. M. Magaro et al., "Influence of Diet with Different Lipid Composition on Neutrophil Chemiluminescence and Disease Activity in Patients with Rheumatoid Arthritis ", Ann Rheu Dis 47 (1988): 793-6.

48. H. Van Der Temple et al., "Effects of Fish Oil Supplementation in Rheumatoid Arthritis ", Ann Rheu Dis 49 (1990): 76-80.

49. J. M. Kremer et al., "Dietary Fish Oil and Olive Oil in Patients with Rheumatoid Arthritis", Arth Rheum 33 (1990): 810-20.

50. C.S Lau, et al., "Maxepa On Non-Steroidal Anti-inflammatory Drug Usage in Patients with Mild Rheumatoid Arthritis ", Br J Rheumatol 30 (1991): 137.

51. G. L. Nielsen et al., "The Effects of Dietary Supplementation with N-3 Polyunsaturated Fatty Acids in Patients with Rheumatoid Arthritis, A Randomized double-blinded trial", Eur J Clin Invest 22 (1992): 678-91.

52. V. Codey, E. Middleton, and J. B. Harborne, Plant Flavonoids in Biology and Medicine- Biochemical, Pharmacological, and Structure-Activity Relationships (New York: Alan R. Liss, 1986); V. Codey, E. Middleton, J. B. Harborne, and A. Beretz, Plant Flavonoids in Biology and Medicine II- Biochemical, Pharmacological, and Structure-Activity Relationships (New York: Alan R. Liss, 1988).

53. U. Tarp at el., "Low Selenium Level in Severe Rheumatoid Arthritis", Scandinavian Journal of Rheumatology 14(1985): 97-101.

54. E. Munthe, and J Aseth, "Treatment of Rheumatoid Arthritis with Selenium and Vitamin E", Scandinavian Journal of Rheumatology 53 (suppl), 1984:103.

55. S.P. Pandley, S.K Bhattacharya, and S. Sundar, "Zinc in Rheumatoid Arthritis ", Indian Journal of Medical Research 81(1985): 618-20.

56. P.A. Simkin, "Treatment of Rheumatoid Arthritis with oral Zinc Sulfate ", Agents and Actions (Suppl.), 8(1981):587-95.

57. P.C. Mattingly, and A.G. Mowat, "Zinc Sulfate in Rheumatoid Arthritis", Annals of the Rheumatic Diseases 41(1982): 456-7.

58. K. B. Menander-Huber, "Orgotein in the Treatment of Rheumatoid Arthritis ", Europ J Rheum inflammation 4(1981): 201-11.

59. S. Zidenberg-Cherr et al., "Dietary Superoxide Dismutase Does not Affect Tissue Level ", Am J Clin Nutr 37(1983): 5-7.
    A. Mullen and C.W.M. Wilson," The Metabolism of Ascorbic Acid in Rheumatoid Arthritis", Proc. Nutr. Sci 35(1976): 8A-9A.

60. N. Subramanian, "Histamine Degradation Potential of Ascorbic Acid ", Agents and Actions 8 (1978): 484-7.

61. M. Levine, "New Concepts in Chemistry and Biochemistry of Ascorbic Acid ", New Engl J Med 314 (1986): 892-902.

62. E.C. Barton-Wright, and W.A Elliott, "The Pantothenic Acid Metabolism of Rheumatoid Arthritis ", Lancet 2 (1963): 862-3.

63. General Practitioner Research Group, "Pantothenic Acid in Rheumatoid Arthritis " : Practitioner 224 (1980): 208-11.

64. B. D. Senturia," Results of Treatment of Chronic Arthritis and Rheumatoid Conditions with Colloidal Sulfur", J Bone Joint Surg 16(1935): 185-8.

65. K. Wheeldon, "The Use of Colloidal Sulfur in the Treatment of Arthritis ", J Bone Joint Surg 17 (1935): 693-726.

66. K. Ransberger, "Enzyme Treatment of Immune Complex Diseases", Arthritis Rheuma 8 (1986) : 16-19.

67. J. A. Shapiro et al., "Diet and Rheumatoid Arthritis in Women: A Possible Protective Effects of Fish Consumption ", Epidemiology 7 (1996): 256-63.

68. G. W. Comstock et al., "Serum Concentrations of Alpha Tocopherol, Beta Carotene, and Retinol Preceding the Diagnosis of Rheumatoid Arthritis and Systematic Lupus Erythematosus ", Ann Rheum Dis 56(1997):323-5.

69. N.F. Childers and G.M. Russo, The Nightshades and Health (Summerville NJ: Horticulture Publications, 1973).

70. T.E. McAlindon et al., "Do Antioxidant Micronutrients Protect against the Development and Progression of Knee Osteoarthritis?", Arthritis Rheumatism 39 (1996):648-56.

71. K. Karzel, and R. Domenjoz, " Effect of Hexosamine Derivatives and Uronic Acid Derivatives on Glycoaminoglycans Metabolism of Fibroblast Cultures", Pharmacology 5 (1971): 337-45.

72. R.R. Vidal y Plana et al., "Articular Cartilage Pharmacology: I. In vitro Studies on Glucosamine and Non-Steroidal Anti-inflammatory Drugs", Pharmacol Res Comm 10 (1978): 557-69.

73. I Setnikar et al., " Pharmacokinetic of Glucosamine in Man", Arzneim Forsch 43 (10) (1993): 1109-13.

A. Biaci et al., Analysis of Glycoaminoglycans in Human Sera after Oral Administration of Chondroitin Sulfate", Rheumatol Int 12 (1992): 81-8.

74. A Conte et al., " Biochemical and Pharmacokinetic Aspect of Oral Treatment with Chondroitin Sulfate", Arzneim Forsch 45 (1995): 918-25.

75. W. Kaufman, The Common Form of Joint Dysfunction: Its Incidence and Treatment (Brattleboro, VT: EL Hildreth Co., 1949.

A. Hoffer, "Treatment of Arthritis by Nicotinic Acid and Nicotinamide", Can Med Assoc J 81(1959): 235-9.

76. K. Lund- Olesen, and K.B. Menander," Orgotein: A new Anti-inflammatory Metalloprotein Drug: Preliminary Evaluation of Clinical Efficacy and Safety in Degenerative Joint Diseases", Curr Ther Res 16 (1974): 706-17.

77. E.C. Huskisson and J. Scott, "Orgotein in Osteoarthritis of the Knee Joint", Eur J Rheumatol Inflam 4 (1981): 212.

I. Machtey, and L. Quaknine, " Toco-phenol in Osteoarthritis: A Controlled Pilot Study", J Am Ger Soc26 (1978): 328-30.

78. E.R. Schwartz, "The Modulation of Osteoarthritic Development by Vitamin C and E", Int J Vit Nutr Res Suppl 26 (1984): 141-46.

79. C.J. Bates, "Proline and Hydroxyproline and Vitamin C Status in Elderly Human Subjects", Clin Sci Mol Med 52 (1977): 535-43.

80. A.P. Prins, J.M. Lipman, C.A.McDevitt, and L. Sokoloff, "Effect of Purified Growth Factors on Rabbit Articular Chondrocytes in Monolayer Culture', Arthr Rheum 25 (1982): 1228-32.

81. J.C. Anand, "Osteoarthritis and Pantothenic Acid", J Coll Gen Pract 5, (1963):136-37.

82. J.C. Anand, "Osteoarthritis and Pantothenic Acid", Lancet 2, (1963):1168.

83. W. Kaufman, The Common Form of Joint Dysfunction: Its Incidence and Treatment (Brattleboro, VT: E.L. Hildreth Company, 1949.

    A. Hoffer, "Treatment of Arthritis by Nicotinic Acid and Nicotinamide", Canadian Medical Association Journal 81 (1959): 235-9.

84. Efthimiou, Petros. (2020). Absolute Rheumatology Review. 10.1007/978-3-030-23022-7.

85. Zhang Y, Peloquin CE, Dubreuil M, et al. Sleep Apnea and the Risk of Incident Gout: A Population-Based, Body Mass Index-Matched Cohort Study. Arthritis Rheumatol. 2015;67(12):3298-3302. PMID: 26477891 doi:10.1002/art.39330

86. Giles TL, et al. (2006). Continuous positive airways pressure for obstructive sleep apnoea in adults. Cochrane Database of Systematic Reviews (3). PMID: 16437429

87. DOI: 10.1002/14651858.CD001106.pub2

88. Neogi T, Chen C, Niu J, et al. Relation of temperature and humidity to the risk of recurrent gout attacks. Am J Epidemiol. 2014;180(4):372-377. DOI: 10.1093/aje/kwu147

89. Ball D., Key J., Introductory Chemistry 1st Canadian Edition, 2014, ISBN: 978-1-77420-002-5.

90. CONCOA Gas Controls, 2012, "Oxygen (O2)", www.concoa.com. Retrieved May 2022.

91. Engineering ToolBox, (2018). *Carbon dioxide - Density and Specific Weight vs. Temperature and Pressure.* [online] Available at: https://www.engineeringtoolbox.com/carbon-dioxide-density-specific-weight-temperature-pressure-d_2018.html. Retrieved May 2022.

92. L. Skoldstam, L. Larsson, and F.D Lindstorm, "Effects of Fasting and Lactovegetarian Diet on Rheumatoid Arthritis", Scand J Rheumatol 8 (1979):137-44.

93. I. Hafstrom, et al., " Effects of Fasting on Disease Activity, Neutrophil Function, Fatty Acid Composition, and Leukotriene Rheumatoid Arthritis", Arthr Rheum 31 (1988): 585-92.

94. R. Petersdorf et al., eds., Harrison's Principles of Internal Medicine (New York: McGraw-Hill, 1983)

95. M.V. Krause and L.K. Mahan, Food, Nutrition, and Diet Therapy, 7th edition, (Philadelphia: W.B Saunders, 1984), 677-9.

96. Neutrition Foundation, Present Knowledge in Nutrition, 5th edition(Washington D.C.: Nutrition Foundation, 1884), 740-56.

97. F.X. Pi-Sunyer, "The Fattening of America", JAMA 272 (1994): 238.

98. R.V. Panganamala and D.G. Cornwell, "the Effect of Vitamin E on Arachidonic Acid Metabolism" Ann NY Acad Sci 393 (1982): 376-91.

99. A.S. Lewis, L. Murphy, C. MCalla, et al., " Inhibition of Mammalian Xanthine Oxidase by Folate Compounds and Amethopterin" , J Biol Chem 259: 15-5, 1984.

100. A. Bindoli, M. Valentine, and L. Cavallini, " Inhibitory Action of Quercetin on Xanthine Oxidase and Xanthine Dehydrogenase Activity", Pharm Res Comm 17(1985): 831-9.

101. S.L. Gershon, and I.H. Fox, " Pharmacological Effects of Nicotinic Acid on Human Purine Metabolism", J Lab Clin Med 84 (1974):179-56.

102. Levick J.R. Hypoxia and acidosis in chronic inflammatory arthritis; relation to vascular supply and dynamic effusion pressure. *J Rheum.* 1990;17(5):579–582.

103. Biniecka M., Kennedy A., Fearon U., Ng C.T., Veale D.J., O'Sullivan J.N. Oxidative damage in synovial tissue is associated with in vivo hypoxic status in the arthritic joint. *Ann Rheum Dis.* 2010;69(6):1172–1178.

Disclaimer

Chapter 1

This chapter is another version of the article published by the same author(s) in the following journal: International Journal of Innovation Scientific Research and Review, 4: 2404-2408, 2022.

Chapter 2

This chapter is another version of the article published by the same author(s) in the following journal: American Journal of Applied and Industrial Chemistry, 6:13-19, 2022.

Chapter 3

This chapter is another version of the article published by the same author(s) in the following journal: American Journal of Applied and Industrial Chemistry, 6(2): 31-35, 2022

Chapter 4

This chapter is another version of the article published by the same author(s) in the following journal: J. Pharmaceutics and Pharmacology Research. 5(8), 2022.

Chapter 5

This chapter is another version of the article published by the same author(s) in the following journal: *Journal of Chemistry, Education Research, and Practice, 5: 127-130, 2021.*

Chapter 6

This chapter is another version of the article published by the same author(s) in the following journal: *Journal of Chemistry, Education Research, and Practice, 5: 127-130, 2021.*

Chapter 7

This chapter is another version of the article published by the same author(s) in the following journal: Scientific Research and Academic Publisher, OJPC. 11, 2021, *DOI: 10.4236/ojpc.2021.112002*

Chapter 8

This chapter is another version of the article published by the same author(s) in the following journal: *American Journal of Physical Chemistry, 10:74-87, 2021.*

Chapter 9

This chapter is another version of the article published by the same author(s) in the following journal: *American Journal of Physical Chemistry*, 11: 32-44, 2022.

Chapter 10

This chapter is another version of the article published by the same author(s) in the following journal: Journal of Pharmacognosy and Phytochemistry, 11(4): 295-302, 2022.

Index

Hydroperoxyl radical of
Hydroperoxide 1- thio-ethyl 60
Hydrogen radical 154
Hydroperoxyl radical of methyl
hydroperoxide 154
Hydroperoxyl radical of ethyl
hydroperoxide 154
Hydroperoxyl radical of propyl
hydroperoxide 154
Hydroperoxide radical 154
Hydrogen peroxide 154
Hydroxide radical 139
Hydroxyl radical of 1,2,2 tri-fluoro
ethanol 157

M
Methyl radical of methyl thio-acetic
acid 60
Methylene radical of methyl thio-
acetic acid 60
Methyl thio-acetate radical 62
Methyl radical of ethyl methyl
sulfide 60
Methyl radical of Hydroperoxide 1-
methyl thioethyl 60
Methylene radical of Hydroperoxide
1-methyl thioethyl 60
Methylene radical of ethyl methyl
sulfide 60
Methyl Sulfanyl Radical of
Hydroperoxide 1-thio-methyl ethyl
60
Methyl Radical of Hydroperoxide 1-
thio-methyl ethyl 60
Methyl fluoride 154
Methyl radical 154
Methyl hydroperoxide 154
Methane 154
Methoxide radical 154

Methanol 154
Methyl radical of 2-fluoro ethanol
72
Methylene radical of 2-fluoro
ethanol 72
Methyl radical of 1-fluoro ethanol
71
Methylene radical of 1-fluoro
ethanol 72
Methylene radical of 1,2 di-fluoro
ethanol 73
Methyl radical of 1,2 di-fluoro
ethanol 73
Methylene radical of 1,1 di-fluoro
ethanol 73
Methyl radical of 2,2 di-fluoro
ethanol 74
Methyl radical of 1,1 di-fluoro
ethanol 74

Methyl sulfanyl radical 139

Methylene radical of 1,1,1 tri-fluoro
ethanol 155
Methyl radical of 1,2,2 tri-fluoro
ethanol 157
Methylene radical of 1,1,2 tri-fluoro
ethanol 156
Methyl radical of 1,1,2 tri-fluoro
ethanol 156
Methylene radical of 1,1,1,2 tetra-
fluoro ethanol 158
Methylene radical of 1,1,1,2 tetra-
fluoro ethanol   158

O
Oxygen radical 154
Oxygen 139

Appendix:
The Periodic Table of Elements

All known elements that make all existing chemical compounds with varying chemical and physical properties discussed in detail in most chapters are arranged by their atomic numbers and listed in the periodic table. The atomic number is the number of protons in the nucleus. The periodic table is the best tool in science, as many chemical and physical properties, can be known based on their position in the periodic table. Periodic tables can be obtained using the internet; a good source of information would be https://ptable.com/#Properties. Most periodic tables contain the atomic mass of each element, which is the number of neutrons and protons in each nucleus. Table 1 lists atomic masses for most elements on the periodic table. Some elements have atomic masses that are known very precisely, to the accuracy of several decimal places; other elements, like radioactive elements, have atomic masses that aren't known precisely; and elements with existing isotopes have varying atomic masses depending on how isotopes are isolated.

Figure 1. The periodic table of elements.

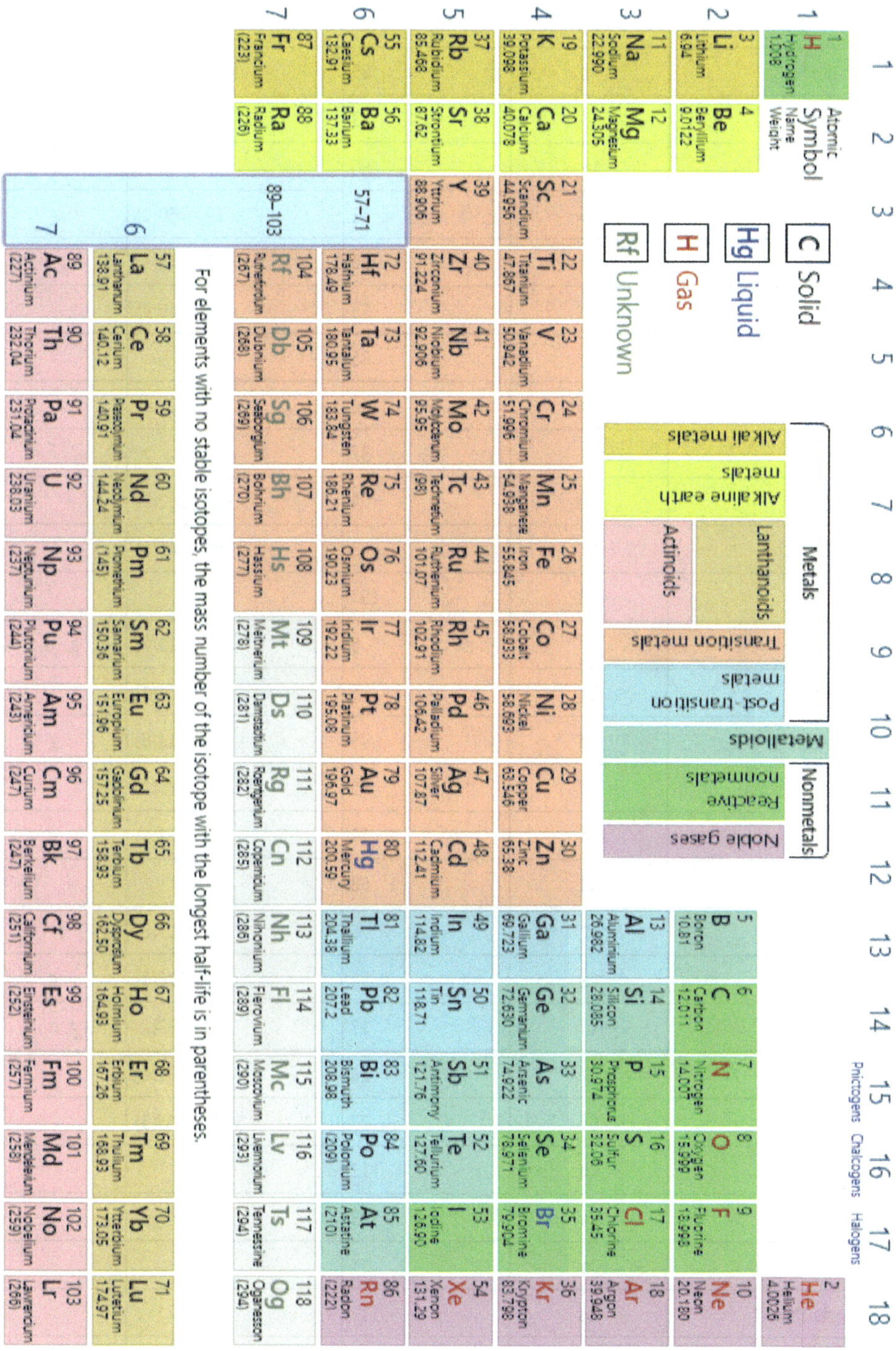

Legend:
- Cell layout: Atomic number (top), Symbol, Name, Weight
- C — Solid
- Hg — Liquid
- H — Gas
- Rf — Unknown

Metals: Alkali metals, Alkaline earth metals, Lanthanoids, Actinoids, Transition metals, Post-transition metals
Metalloids
Nonmetals: Reactive nonmetals, Noble gases

Group 15 — Pnictogens; Group 16 — Chalcogens; Group 17 — Halogens.

1	2	3	4	5	6	7	8	9	10	11	12	13	14	15	16	17	18
1 H Hydrogen 1.008																	2 He Helium 4.0026
3 Li Lithium 6.94	4 Be Beryllium 9.0122											5 B Boron 10.81	6 C Carbon 12.011	7 N Nitrogen 14.007	8 O Oxygen 15.999	9 F Fluorine 18.998	10 Ne Neon 20.180
11 Na Sodium 22.990	12 Mg Magnesium 24.305											13 Al Aluminium 26.982	14 Si Silicon 28.085	15 P Phosphorus 30.974	16 S Sulfur 32.06	17 Cl Chlorine 35.45	18 Ar Argon 39.948
19 K Potassium 39.098	20 Ca Calcium 40.078	21 Sc Scandium 44.956	22 Ti Titanium 47.867	23 V Vanadium 50.942	24 Cr Chromium 51.996	25 Mn Manganese 54.938	26 Fe Iron 55.845	27 Co Cobalt 58.933	28 Ni Nickel 58.693	29 Cu Copper 63.546	30 Zn Zinc 65.38	31 Ga Gallium 69.723	32 Ge Germanium 72.630	33 As Arsenic 74.922	34 Se Selenium 78.971	35 Br Bromine 79.904	36 Kr Krypton 83.798
37 Rb Rubidium 85.468	38 Sr Strontium 87.62	39 Y Yttrium 88.906	40 Zr Zirconium 91.224	41 Nb Niobium 92.906	42 Mo Molybdenum 95.95	43 Tc Technetium (98)	44 Ru Ruthenium 101.07	45 Rh Rhodium 102.91	46 Pd Palladium 106.42	47 Ag Silver 107.87	48 Cd Cadmium 112.41	49 In Indium 114.82	50 Sn Tin 118.71	51 Sb Antimony 121.76	52 Te Tellurium 127.60	53 I Iodine 126.90	54 Xe Xenon 131.29
55 Cs Caesium 132.91	56 Ba Barium 137.33	57–71	72 Hf Hafnium 178.49	73 Ta Tantalum 180.95	74 W Tungsten 183.84	75 Re Rhenium 186.21	76 Os Osmium 190.23	77 Ir Iridium 192.22	78 Pt Platinum 195.08	79 Au Gold 196.97	80 Hg Mercury 200.59	81 Tl Thallium 204.38	82 Pb Lead 207.2	83 Bi Bismuth 208.98	84 Po Polonium (209)	85 At Astatine (210)	86 Rn Radon (222)
87 Fr Francium (223)	88 Ra Radium (226)	89–103	104 Rf Rutherfordium (267)	105 Db Dubnium (268)	106 Sg Seaborgium (269)	107 Bh Bohrium (270)	108 Hs Hassium (277)	109 Mt Meitnerium (278)	110 Ds Darmstadtium (281)	111 Rg Roentgenium (282)	112 Cn Copernicium (285)	113 Nh Nihonium (286)	114 Fl Flerovium (289)	115 Mc Moscovium (290)	116 Lv Livermorium (293)	117 Ts Tennessine (294)	118 Og Oganesson (294)

For elements with no stable isotopes, the mass number of the isotope with the longest half-life is in parentheses.

Lanthanoids (period 6):

57 La Lanthanum 138.91	58 Ce Cerium 140.12	59 Pr Praseodymium 140.91	60 Nd Neodymium 144.24	61 Pm Promethium (145)	62 Sm Samarium 150.36	63 Eu Europium 151.96	64 Gd Gadolinium 157.25	65 Tb Terbium 158.93	66 Dy Dysprosium 162.50	67 Ho Holmium 164.93	68 Er Erbium 167.26	69 Tm Thulium 168.93	70 Yb Ytterbium 173.05	71 Lu Lutetium 174.97

Actinoids (period 7):

89 Ac Actinium (227)	90 Th Thorium 232.04	91 Pa Protactinium 231.04	92 U Uranium 238.03	93 Np Neptunium (237)	94 Pu Plutonium (244)	95 Am Americium (243)	96 Cm Curium (247)	97 Bk Berkelium (247)	98 Cf Californium (251)	99 Es Einsteinium (252)	100 Fm Fermium (257)	101 Md Mendelevium (258)	102 No Nobelium (259)	103 Lr Lawrencium (266)

Table 1. Common Information of Elements of the Periodic Table

Name	Atomic Symbol	Atomic Number	Atomic Mass	Footnotes
actinium*	Ac	89		
aluminum	Al	13	26.9815386(8)	
americium*	Am	95		
antimony	Sb	51	121.760(1)	g
argon	Ar	18	39.948(1)	g, r
arsenic	As	33	74.92160(2)	
astatine*	At	85		
barium	Ba	56	137.327(7)	
berkelium*	Bk	97		
beryllium	Be	4	9.012182(3)	
bismuth	Bi	83	208.98040(1)	
bohrium*	Bh	107		
boron	B	5	10.811(7)	g, m, r
bromine	Br	35	79.904(1)	
cadmium	Cd	48	112.411(8)	g
caesium (cesium)	Cs	55	132.9054519(2)	
calcium	Ca	20	40.078(4)	g
californium*	Cf	98		
carbon	C	6	12.0107(8)	g, r
cerium	Ce	58	140.116(1)	g
chlorine	Cl	17	35.453(2)	g, m, r
chromium	Cr	24	51.9961(6)	
cobalt	Co	27	58.933195(5)	
copernicium*	Cn	112		
copper	Cu	29	63.546(3)	r
curium*	Cm	96		
darmstadtium*	Ds	110		
dubnium*	Db	105		

dysprosium	Dy	66	162.500(1)	g
einsteinium*	Es	99		
erbium	Er	68	167.259(3)	g
europium	Eu	63	151.964(1)	g
fermium*	Fm	100		
fluorine	F	9	18.9984032(5)	
francium*	Fr	87		
gadolinium	Gd	64	157.25(3)	g
gallium	Ga	31	69.723(1)	
germanium	Ge	32	72.64(1)	
gold	Au	79	196.966569(4)	
hafnium	Hf	72	178.49(2)	
hassium*	Hs	108		
helium	He	2	4.002602(2)	g, r
holmium	Ho	67	164.93032(2)	
hydrogen	H	1	1.00794(7)	g, m, r
indium	In	49	114.818(3)	
iodine	I	53	126.90447(3)	
iridium	Ir	77	192.217(3)	
iron	Fe	26	55.845(2)	
krypton	Kr	36	83.798(2)	g, m
lanthanum	La	57	138.90547(7)	g
lawrencium*	Lr	103		
lead	Pb	82	207.2(1)	g, r
lithium	Li	3	[6.941(2)]†	g, m, r
lutetium	Lu	71	174.967(1)	g
magnesium	Mg	12	24.3050(6)	
manganese	Mn	25	54.938045(5)	
meitnerium*	Mt	109		
mendelevium*	Md	101		

mercury	Hg	80	200.59(2)	
molybdenum	Mo	42	95.94(2)	g
neodymium	Nd	60	144.242(3)	g
neon	Ne	10	20.1797(6)	g, m
neptunium*	Np	93		
nickel	Ni	28	58.6934(2)	
niobium	Nb	41	92.90638(2)	
nitrogen	N	7	14.0067(2)	g, r
nobelium*	No	102		
osmium	Os	76	190.23(3)	g
oxygen	O	8	15.9994(3)	g, r
palladium	Pd	46	106.42(1)	g
phosphorus	P	15	30.973762(2)	
platinum	Pt	78	195.084(9)	
plutonium*	Pu	94		
polonium*	Po	84		
potassium	K	19	39.0983(1)	
praseodymium	Pr	59	140.90765(2)	
promethium*	Pm	61		
protactinium*	Pa	91	231.03588(2)	
radium*	Ra	88		
radon*	Rn	86		
roentgenium*	Rg	111		
rhenium	Re	75	186.207(1)	
rhodium	Rh	45	102.90550(2)	
rubidium	Rb	37	85.4678(3)	g
ruthenium	Ru	44	101.07(2)	g
rutherfordium*	Rf	104		
samarium	Sm	62	150.36(2)	g

scandium	Sc	21	44.955912(6)	
seaborgium*	Sg	106		
selenium	Se	34	78.96(3)	r
silicon	Si	14	28.0855(3)	r
silver	Ag	47	107.8682(2)	g
sodium	Na	11	22.98976928(2)	
strontium	Sr	38	87.62(1)	g, r
sulfur	S	16	32.065(5)	g, r
tantalum	Ta	73	180.94788(2)	
technetium*	Tc	43		
tellurium	Te	52	127.60(3)	g
terbium	Tb	65	158.92535(2)	
thallium	Tl	81	204.3833(2)	
thorium*	Th	90	232.03806(2)	g
thulium	Tm	69	168.93421(2)	
tin	Sn	50	118.710(7)	g
titanium	Ti	22	47.867(1)	
tungsten	W	74	183.84(1)	
ununhexium*	Uuh	116		
ununoctium*	Uuo	118		
ununpentium*	Uup	115		
ununquadium*	Uuq	114		
ununtrium*	Uut	113		
uranium*	U	92	238.02891(3)	g, m
vanadium	V	23	50.9415(1)	
xenon	Xe	54	131.293(6)	g, m
ytterbium	Yb	70	173.04(3)	g
yttrium	Y	39	88.90585(2)	
zinc	Zn	30	65.409(4)	

| zirconium | Zr | 40 | 91.224(2) | g |

*Element has no stable nuclides. However, three such elements (Th, Pa, and U) have a characteristic terrestrial isotopic composition, and for these an atomic mass is tabulated.

†Commercially available Li materials have atomic weights that range between 6.939 and 6.996; if a more accurate value is required, it must be determined for the specific material.

g Geological specimens are known in which the element has an isotopic composition outside the limits for normal material. The difference between the atomic mass of the element in such specimens and that given in the table may exceed the stated uncertainty.

m Modified isotopic compositions may be found in commercially available material because it has been subjected to an undisclosed or inadvertent isotopic fractionation. Substantial deviations in the atomic mass of the element from that given in the table can occur.

r Range in isotopic composition of normal terrestrial material prevents a more precise $Ar(E)$ being given; the tabulated $Ar(E)$ value and uncertainty should be applicable to normal material.

Source: Adapted from *Pure and Applied Chemistry* 78, no. 11 (2005): 2051–66. © IUPAC (International Union of Pure and Applied Chemistry).

Made in the USA
Monee, IL
07 July 2026

56552469R00136